Military Decision-Making Processes

ALSO BY KEVIN DOUGHERTY

The United States Military in Limited War: Case Studies in Success and Failure, 1945–1999 (McFarland, 2012)

Military Decision-Making Processes

*Case Studies Involving the Preparation,
Commitment, Application
and Withdrawal of Force*

KEVIN DOUGHERTY

McFarland & Company, Inc., Publishers
Jefferson, North Carolina, and London

LIBRARY OF CONGRESS CATALOGUING-IN-PUBLICATION DATA

Dougherty, Kevin.
Military decision-making processes : case studies involving the preparation, commitment, application and withdrawal of force / Kevin Dougherty.
pages   cm.
Includes bibliographical references and index.

ISBN 978-0-7864-7798-2
softcover : acid free paper

1. United States—Military policy—Decision making—Case studies.
2. National security—United States—Decision making—Case studies.
3. United States—Foreign relations—Decision making—Case studies.
4. Strategy—Case studies.   5. United States—History, Military.   I. Title.
UA23.D6923 2014   355.6'830973—dc23   2013036290

BRITISH LIBRARY CATALOGUING DATA ARE AVAILABLE

On the cover (top to bottom): The *Enola Gay* landing after its bomb run (USAF); a mission map for the bombing of Hiroshima; the mushroom cloud over the city (USAF); and a map of the blast and fire damage (U.S. Strategic Bombing Survey)

Manufactured in the United States of America

McFarland & Company, Inc., Publishers
Box 611, Jefferson, North Carolina 28640
www.mcfarlandpub.com

This book is lovingly dedicated to my children,
Zack and Kathryn,
in the hope they will always make good decisions.

Contents

Preface

President Bill Clinton no doubt spoke for all his fellow commanders-in-chief when in 1995, on the eve of his decision to deploy ground troops to Bosnia, he declared he had "no responsibility more grave than putting soldiers in harm's way."[1] I was about halfway through my Army career at the time, and I was a small part of the cast of hundreds at Headquarters, United States Army Europe involved in the planning of what would become the Implementation Force that would deploy to Bosnia in pursuit of President Clinton's policy. Like all Army officers, I was school-trained in the rational actor model of decision-making. Furthermore, I was also a product of an Army that largely embraced Secretary of Defense Caspar Weinberger's rather restrictive criteria for the use of force. Even as a junior Army major, though, I could see that the world and the role the US military would play in it was changing, and that these comfortable touchstones would no longer be sufficient. The more I learned, the more I realized that there would be, and always had been, alternatives.

President Clinton's statement suggests that a study of the decision-making process associated with the weighty matters concerning the use of force would be both interesting and enlightening. Indeed, it is. It does not take even the most casual observer of the subject long, however, to realize the process is neither standardized nor simple. While all individuals associated with important decisions about national security and the lives of America's service members take their responsibilities seriously, the processes by which they reach their conclusions are varied and complicated. The phenomenon had always intrigued me on some level, but the depth of its chimeric nature became more apparent to me as the United States pondered the decision of what to do about the civil unrest in Libya in 2011. When one observer described the decision to support the anti–Qaddafi rebels with air strikes as a "just enough doctrine," designed to take "only those steps that are likely to produce a satisfactory outcome, rather than guaranteeing an optimal one," I was somewhat confused.[2] Why, I wondered, would the United States settle for something suboptimal, with all that implied?

As I pursued the answer, I consulted many previous works on the subject. Several made detailed and compelling cases by focusing on a single decision-making model. In *Understanding Foreign Policy Decision Making* (Cambridge, 2010), for example, Alex Mintz and Karl DeRouen approach the problem from the vantage point of poliheuristic theory.

Likewise, in *Risk-taking in International Politics* (University of Michigan Press, 2001), Rose McDermott uses prospect theory as the basis of her analysis. Other authors analyzed decisions based on a certain set of characteristics. In *Decisions and Dilemmas: Case Studies in Presidential Foreign Policy Making Since 1945* (M. E. Sharpe, 2005), for example, Robert Strong explains his cases in terms of power, process, and personality.

As I surveyed the literature, I decided the contribution I might be able to make was to equally and objectively introduce and illustrate a number of different decision-making models. I had learned that no one model applied in every case, and the discerning observer was best served by a familiarity with an array of choices which could then be matched to an appropriate situation. Under the same assumption that the variety of explanations required a broad treatment, I endeavored to expand the cases beyond the limited numbers and strategic focus I had found in other volumes. Thus, *Military Decision-Making Processes: Case Studies Involving the Preparation, Commitment, Application and Withdrawal of Force* includes eight decision-making models illustrated by twenty-four diverse cases. Because of this general and inclusive philosophy, the book takes an interdisciplinary approach that is intended to be of interest to readers concerned with political science and public administration, military history and international relations, and decision-making and leadership. In the process, it aims to be detailed enough for scholarly, and engaging enough for popular, audiences.

Introduction

Everyone makes decisions and is affected by decisions made by others. Among the most interesting and impactful decisions made in the United States are those involving the use of force. This book uses an eclectic selection of such decisions to illustrate various decision-making models and their relation to the American way of war.

The book traces traditional and emerging theories of decision-making by first explaining the components of the model and then analyzing its practical application through three case studies. Each chapter concludes with a discussion of the utility and explanatory power of the particular theory. Because even at their very best, a particular decision-making theory can only explain some cases, the chapter then segues to another theory that has different characteristics.

The book begins with the rational actor model which portrays decisions as being the result of a systematic and logical process designed to produce the optimal outcome in terms of an agreed upon set of evaluation criteria. The rational actor model is useful in explaining President Harry Truman's decision to use atomic weapons in World War II, President Ronald Reagan's decision to bomb Libya in 1986, and President George Bush's decision to cease offensive operations in Operation Desert Storm. As intuitively appealing as the rational approach may be, its critics point out that it simply does not reflect the way all decisions are made in actual practice. Instead, they argue the relevance of cognitive approaches that reflect the importance of psychological factors. The prospect theory is one such model.

Prospect theory posits that the domain of the decision-maker is central to the process. Decision-makers who perceive themselves as in the domain of gains tend to be risk-averse while those in the domain of losses tend to be risk-acceptant. Prospect theory is useful in explaining General George Washington's decision to attack the British at Trenton in 1776, Major General George Meade's decision not to aggressively pursue General Robert E. Lee after the Battle of Gettysburg in 1863, and President Bill Clinton's decision to withdraw American forces from Somalia after the Battle of Mogadishu in 1993. However, even prospect theory's proponents admit that decision-making is a complex process that is best understood through the interaction of several models. Poliheuristic theory adopts this theme by explaining decision-making as both a cognitive and rational process.

Poliheuristic theory holds that decisions are made as a result of a two-stage process. First, the decision-maker simplifies the problem by using cognitive short-cuts or heuristics. This quick first cut eliminates any option that involves any unacceptable political costs. Then the decision-maker evaluates the remaining alternatives using a detailed, rational cost-benefit analysis. Poliheuristic theory is useful in explaining President Dwight Eisenhower's decision not to intervene in Hungary in 1956, President John Kennedy's authorization of the Bay of Pigs invasion in 1961, and President Lyndon Johnson's decision to deescalate American involvement in Vietnam after the Tet Offensive in 1968. However, poliheuristic theory has been criticized for appearing to be limited to decisions in which a single predominant leader has the ultimate authority to choose a course of action. The bureaucratic model may be more useful in accounting for the importance of organizations in the decision-making process.

The bureaucratic model argues that decisions result from the bargaining process among various government agencies that have somewhat different interests in the outcome. It is useful in explaining the impact of the Confederate departmental system at Vicksburg in the Civil War, the battle between the Army and Air Force over helicopter fielding in the 1950s and 1960s, and the roles of the Department of State and the Department of Defense in the decision to commit a peacekeeping force to Beirut in 1982. However, the bureaucratic model is most applicable when time is available for political bargaining. It is of less usefulness in those countless cases where decisions are more routinely and rapidly made on the basis of some *a priori* guideline or administrative rule. In such instances, the organizational process model may be a greater analytical tool.

The organizational process model recognizes that many decisions are made within agencies by routines or standard operating procedures that organizations have developed over time. It is useful in explaining the formulation of American strategy during the Vietnam War, the decision to abort the Iranian hostage rescue mission in 1980, and the federalization of the California Army National Guard during the Los Angeles Riots in 1992. However, if the established procedure is too rigid to meet the needs of the situation or the situation is one with which the organization has no significant prior experience, a small group may be formed to make the decision.

The small group model recognizes that many decisions are made by neither individuals nor organizations, but by small groups convened for a specific purpose. It is useful in explaining the Blockade Board convened in 1861 to develop a Union blockade strategy, the Executive Committee formed in response to the Cuban Missile Crisis in 1962, and the Restricted Interagency Group which was tasked with developing a strategy toward Nicaragua in the 1980s. Small groups are formed by a higher authority such as the president or a cabinet secretary and operate at his direction. Sometimes, however, actors outside the hierarchical decision-making process are able to exert a disproportionate influence on it. The elite theory is concerned with this phenomenon.

Elite theory gives high priority to the influence of the identity of the individuals making decisions and their ability to manipulate the underlying dynamics of national power, social myth, and class interests to serve the interests of a small sector of society. It is useful in explaining the role of the media in the decision to intervene in Somalia in 1992, the influence of the Congressional Black Caucus in the decision to intervene in Haiti in 1993, and the agenda of Secretary of State Madeline Albright in the decision to intervene in Kosovo in 1999. The elite theory recognizes that power resources are not evenly distributed throughout society and that many groups in society have power to participate in policy making. The pluralism model builds on these understandings, but in a way that ideally is more compatible with democratic principles.

Pluralism argues that merely possessing the attributes of power does not necessarily equate to actually possessing power itself and that no one group is powerful enough to dictate policy. Thus decisions result from bargaining between groups and reflect the interest of the dominant group(s). Pluralism is useful in explaining conscription in the Confederate Army, the decision to reflag Kuwaiti tankers in the midst of the Iran–Iraq War, and the nature of the Implementation Force committed to Bosnia in 1995.

Just as pluralism recognizes that no one group has the power to make all decisions, no one model has the power to explain every decision. Indeed with a little imagination, many decisions can be explained by several models simultaneously. For example, rather than the rational actor model used here, Barton Bernstein implies a poliheuristic explanation of President Harry Truman's use the atomic bomb in arguing that Truman inherited from President Franklin Roosevelt the commitment to use the bomb as a combat weapon and was left only with the remaining decision of "how to fulfill" this legacy Roosevelt had left him.[1] The Iranian Hostage Rescue mission is used to illustrate the organizational process model in this volume whereas elsewhere Rose McDermott skillfully presents it from another angle as an example of the prospect theory.[2] Here, the Executive Committee is offered as an example of small group decision-making during the Cuban Missile Crisis. In his seminal "Conceptual Models and the Cuban Missile Crisis," Graham Allison explores the event using the rational actor, organizational process, and bureaucratic models.[3] This present survey makes no claim to the exclusivity of the models selected to analyze its cases. Rather, it is designed to present some of the more common theories and illustrate them in the context of a decision involving the use of force. Armed with these examples, the discerning reader will no doubt be able to apply the models to a variety of other instances.

1

The Rational Actor Model

For many observers, the rational actor model represents "a common starting point for studying the decision-making process."[1] As its name suggests, the theory is based on a rationality which Graham Allison defines as a "consistent, value-maximizing choice within specified constraints."[2] Thus a decision-maker following this methodology can be expected to employ purposive action, display consistent preferences, and seek to maximize utility.[3] Purposive action for the rational actor results from goal-oriented motivation and behavior rather than mere habit or social expectations. The model assumes that at least at a general level, decision-makers share a common vision of the national interest and a unity of purpose that allows them to identify an *a priori* goal and collectively move toward achievement of that goal. Consistency manifests itself both in transitivity and invariance. Transitivity requires that in a relationship among three entities, if there is a relationship between the first and second entities and between the second and third entities, the relationship also holds between the first and third entities. Thus if the rational decision-maker prefers Outcome 1 to Outcome 2, and Outcome 2 to Outcome 3, he necessarily prefers Outcome 1 to Outcome 3. Invariance means that the decision-maker's preferences will remain unchanged regardless of how the information is presented. Utility maximization occurs when decision-makers systematically consider the entire range of alternatives and select the one that will most efficiently and effectively promote the national interests.[4] The decision to use the atomic bomb in World War II, the bombing of Libya after the La Belle Discotheque terrorist attack, and the decision to end Operation Desert Storm all illustrate the rational actor model.

Example 1: Truman's Decision
to Use the Atomic Bomb on Japan, 1945

Greg Cashman describes the rational actor decision-making process as consisting of the following steps:

1. Identify and define the problem.
2. Identify goals.
3. Gather information.

4. Analyze each possible alternative.
5. Choose the alternative best able to achieve your objectives.
6. Implement the decision.
7. Monitor and evaluate your policy decision.
8. Terminate/alter/continue the policy as determined by your evaluation of it in Step 7.[5]

President Harry Truman's decision to drop the atomic bomb on Hiroshima, Japan, on August 6, 1945, and then again on Nagasaki on August 9 reflects this sequence. Indeed, Truman's popular image as the champion of "the buck stops here" certainly is consistent with the rational actor methodology.[6]

Following Cashman's procedure, the problem facing Truman in 1945 was how to gain a Japanese surrender. This problem is not the same as how to defeat Japan, which Truman's biographer David McCullough convincingly explains "was not the issue."[7] By July, the tide of battle had definitely turned in the Allies' favor. Japan's economy was "disintegrating rapidly."[8] General Dwight Eisenhower seemed to concur with intelligence reports that "indicated the imminence of Japan's collapse."[9] Even internal Japanese studies had concluded by January 1944 that the war was lost. But inevitable defeat and immediate surrender were not the same. The Japanese continued to fight on, inflicting nearly half the total American battle casualties from three years of fighting in the Pacific in the three months since Truman had taken office. "The nearer victory came, the heavier price in blood," writes McCullough.[10] Truman understood the urgency of the situation. "Every day peace is delayed," he told a joint session of Congress on April 16, "costs a terrible toll."[11] The problem as Truman defined it was Japanese surrender, and fast.

Once the problem is identified, there are likely to be numerous possible solutions. The rational actor model helps the decision-maker manage this set by employing expected utility theory. The decision-maker does not merely want a solution to the problem. He wants the one that generates the greatest degree of satisfaction of specific goals and objectives situated within the solution.[12] These goals are identified in Step 2 of Cashman's process. For Truman, the ability to "shorten the agony of war" and in the process "save the lives of thousands and thousands of young Americans" became the goals associated with any potential solution to the problem.[13]

Gathering information was a particularly challenging step for Truman given the secrecy surrounding the bomb. He assumed office on April 12 after the death of President Franklin Roosevelt, and it was not until that day that Secretary of War Henry Stimson briefly informed the new president that a project was underway to produce an atomic bomb. Truman recorded that this small exchange "was the first bit of information that had come to me about the atomic bomb, but [Stimson] gave me no details."[14] On April 24, Stimson requested a meeting to further discuss the issue, and the next day gave Truman his first detailed account of what was known as the Manhattan Project.[15]

This effort was brought together in 1942 and took its name from the location of its first office. Its catalyst was a group of émigré scientists who believed German physicists were making important progress in the area of nuclear fission that could ultimately lead to the development of an atomic bomb. Fearful of the possession of such a powerful weapon in Nazi hands, the group convinced Albert Einstein to transmit their concerns to President Roosevelt. Einstein did so in a letter in October 1939, and Roosevelt was sufficiently impressed to authorize a coordinating commission and later diverted mili-

The USS *Bunker Hill* ablaze after being struck by two kamikaze planes off Okinawa on May 11, 1945 (Naval Historical Center).

tary funds to the top-secret research. In September 1942, US Army engineer Leslie Groves was appointed to the rank of brigadier general and assigned to direct the project.[16]

Groves assembled a staff of scientific, technical, and administrative experts and eventually supervised a force of over 125,000 people and a $2 billion budget. To facilitate secrecy, the work was highly compartmentalized and dispersed throughout the country. Major facilities were located in Los Alamos, New Mexico; Oak Ridge, Tennessee; and Hanford, Washington. Among the leading scientists working on the project were Robert Oppenheimer, Enrico Fermi, and Leo Szilad.[17]

From the inception of the Manhattan Project, most policy makers assumed the bomb to be a legitimate weapon and one that would be used. They also understood the urgency of the situation. In October 1942, for example, Secretary of War Stimson told Brigadier General Groves that the mission was "to produce [the bomb] at the earliest possible date so as to bring the war to a conclusion." Whenever "a single day could be saved," Groves was told to do so.[18]

In spite of this emphasis and effort, there were only three atomic bombs under construction in the summer of 1945. One was successfully detonated on July 16 at Alamogordo, New Mexico, but even then there was uncertainty whether the other two bombs would work.[19] Even if they did explode, Truman still "did not know as yet what effect the new weapon might have, physically or psychologically, when used against the enemy."[20]

Of greater certainty was that a ground invasion of the Japanese home islands would be extremely costly. The invasion of Okinawa in May had resulted in 49,151 American battle casualties; more than in any previous campaign against the Japanese. The fierce fighting left 12,520 Americans killed or missing and 36,631 wounded. Non-battle casualties during the campaign included 15,613 for the Army and 10,598 for the Marines. Thirty-six ships were sunk and 368 damaged. Between April 1 and July 1,763 planes were lost. The Japanese resistance had been fanatical with the enemy suffering approximately 110,000 dead. Tellingly, only 7,400 were taken prisoners, indicating the Japanese willingness to fight to the death.[21] As Arthur Cyr notes, "contrary to historical experience that a nation's will to fight erodes as defeat nears, the Japanese military seemed to fight ever harder as the US forces approached the home islands of Japan."[22]

Based on this information, Truman had two alternatives: use the unprecedented atomic bomb with all its implications of introducing to the world a new era of destruction or launch a conventional invasion of the Japanese home islands. He did not consider a technical demonstration such as an explosion over a deserted island as a possible course of action because it would not "be likely to bring the war to an end."[23] Such methodology is consistent with the rational actor model's demand that to be considered, alternatives must be able to achieve the objective.[24] In Step 4, the president analyzed these two choices in an effort to optimize his goals of hastening the war's conclusion and minimizing American casualties. There were of course other variables to consider. Secretary Stimson, for example, had left Truman with the impression he was "at least as much concerned with the role of the atomic bomb in the shaping of history as in its capacity to shorten this war."[25] In such cases, the rational actor model demands a ranking of goals in order of importance.[26]

For President Truman, clearly the priority was ending the war quickly in a way that saved American lives, and the information he had gathered in Step 3 supported the idea that using the atomic bomb best achieved these goals. Stimson had concluded that "to extract a genuine surrender from the Emperor and his military advisors, there must be administered a tremendous shock which could carry convincing proof of our power to destroy the Empire."[27] If Truman opted for an invasion of the Japanese home islands, General George Marshall estimated that it could cost half a million American lives to compel a surrender.[28] Such figures were hard to quantify, but as McCullough explains, "whatever the projected toll in American lives in an invasion, it was too high if it could be avoided."[29] Using the atomic bomb might unleash some long-range problems, but these were less important to Truman whose principal goal was to end the war.[30] Also less important were any potential benefits in shaping the post-war world by impressing the Soviet Union with the use of the bomb. Barton Bernstein correctly analyzes this relationship, explaining that a more tractable Soviet Union was "an additional reason reinforcing an earlier analysis," but "ending the war speedily was the primary purpose" behind Truman's decision.[31] Indeed, the president ultimately explained to the American people, "We have used [the atomic bomb] in order to shorten the agony of war, in order to save the lives of thousands and thousands of young Americans."[32] Marshall agreed, reflecting the same prioritization in saying, "We had to end the war; we had to save American lives."[33]

As a result of this analysis, Truman opted in Step 5 to employ the atomic bomb. He emphasizes in his *Memoirs* that the ultimate responsibility for this momentous decision rested with him alone, but Robert Strong also points out that the president's decision was the result of a "process ... [that] involved serious policy makers in careful deliberations."[34] Stimson chaired an august group called the Interim Committee to make recommendations on a host of issues including use of the atomic bomb. On June 1, the committee concluded, "the present view of the Committee was that the bomb should be used against Japan as soon as possible, that it be used on a war plant surrounded by workers' homes; and that it be used without prior warning."[35]

Although the rational actor model assumes a shared sense of national interest, it does not require total agreement among the participants in every step of the decision-making process.[36] Indeed, there was debate among the president's advisors about such issues as giving warning and target selection. Assistant Secretary of War John McCloy, for example, suggested some diplomatic overtures as options, and Interim Committee member Ralph Bard argued for giving warning.[37] However, such alternatives were all considered as part of an orderly process that focused on solving the defined problem in the context of the identified goals. No competing interest emerged as a viable threat to the general consensus. Indeed, even the scientific community associated with the Manhattan Project made "no claim to special competence in solving the political, social, and military problems which are presented by the advent of atomic power."[38] As Strong concludes, "a war-weary nation, focused on bringing its conflict with Japan to an end as rapidly as possible, used all available means to accomplish that end."[39]

In Step 6, President Truman implemented his decision. On July 24, he ordered General Carl Spatz, Commanding General of the United States Army Strategic Air Forces, "to deliver its first special bomb as soon as weather will permit visual bombing after about 3 August."[40] When Japan failed to surrender by August 3, the order was carried out and the crew of the *Enola Gay* dropped an atomic bomb on Hiroshima on August 6. In spite of suffering some 130,000 casualties, the Japanese government did not surrender.[41]

A wave of scholarship, often in the revisionist category, has sought to explain Japan's unwillingness to surrender even after the bombing of Hiroshima. A common theme is that, given the destruction of Japan's communications system, the United States simply did not allow the enemy sufficient time to process and respond to the development.[42] For whatever the reason, the Hiroshima bombing did not significantly change Japan's policy.[43] Given that his objective had not been met, President Truman thus moved on to Cashman's Step 8 and had to decide whether to terminate, alter, or continue the policy.

President Truman was prepared for such an eventuality. When he first announced the bombing of Hiroshima, he noted that the target had been selected because it was "a military base" and "we wished in this first attack to avoid, insofar as possible, the killing of civilians." He added, however, "that attack is only a warning of things to come. If Japan does not surrender, bombs will have to be dropped on her war industries and, unfortunately, thousands of civilian lives will be lost."[44] Truman concluded, "We shall continue to use [the atomic bomb] until we completely destroy Japan's power to make war. Only a Japanese surrender will stop us."[45]

After the bombing of Hiroshima, President Truman continued to signal Japan that "we meant business" with strong B-29 attacks on August 7 and 8. Then on August 9, the United States dropped a second atomic bomb; this time on Nagasaki. Truman reports, "This second demonstration of the power of the atomic bomb apparently threw Tokyo into a panic, for the next morning brought the first indication that the Japanese Empire was ready to surrender."[46] Observers debate exactly why Japan decided to surrender at this particular point. Tsuyoshi Hasegawa, for example, argues the decision was prompted more by the Soviet declaration of war than the bombing of Nagasaki.[47] Such discussion, however, focuses on Japanese decision-making and should not be confused with this study of American decision-making. Clearly, Truman, in applying the rational actor model, assessed the situation and decided to continue his policy using atomic weapons to pursue his objective of compelling Japan to surrender.

Example 2: Reagan's Decision to Bomb Libya, 1986

In explaining the rational actor model, Russell Bova notes that "foreign policy-making is all about objectively linking ends and means."[48] "Ends" are the objectives that must be accomplished to advance the national interests. "Means" are resources and elements of power that the nation has available to achieve those objectives.[49] "Ways" are what strategists call the concepts that describe how the means will be used to achieve the ends.[50] So in the rational actor model, the strategist must first determine what he wants to get done and then what he can use and how he can do it.

The beginning of the rational actor decision-making process is to determine the objective. When the Vietnam War exposed America's vulnerability, a wave of terrorism that David Rapoport describes as the "New Left" began to emerge to challenge the existing system. Terrorism resumed an international dimension with a revolutionary ethos creating significant bonds between separate national groups. Many Western groups including the American Weather Underground, the West German Red Army Faction, the Italian Red Brigades, the Japanese Red Army, and the French *Action Directe* "saw themselves as vanguards for the Third World masses." In the Middle East, the Palestinian Liberation Organization reached the heroic status the Viet Cong had garnered in Vietnam. The Soviet Union encouraged terrorism and supported it with training and equipment as part of the Cold War struggle.[51] Bruce Jentleson notes that "by early 1986 terrorism had become 'a growth industry.'"[52]

One-third of the international New Left terrorist attacks involved American targets.[53] United States Navy diver Robert Dean Stethem was beaten and killed during the June 1985 hijacking of TWA Flight 847, and his body dumped on the tarmac at the Beirut airport. The elderly and wheelchair-bound Jewish-American Leon Klinghoffer was shot and thrown overboard from the *Achille Lauro* when the ship was hijacked in October 1985. An eleven year-old girl was killed and several other Americans were wounded during twin attacks at the Rome and Vienna airports on December 27, 1985. The La Belle discotheque bombing in Berlin in April 1986 injured 230 people and killed two, including an American soldier.[54]

Libya was considered to be heavily involved in orchestrating many of these and other acts of international terrorism. In February 1986, the country helped finance an international terrorism congress held in Frankfurt, Germany, that met under the slogan, "The armed struggle as a strategic and tactical necessity in the fight for revolution." As many as 500 people were reported to have attended the congress, including representatives from German, French, Belgian, Spanish, and Portuguese terrorists, as well as members of the PLO, the Popular Front for the Liberation of Palestine, the African National Congress, the Irish Republican Army, the Tupamaros, the Italian Red Brigades, and the Basque ETA. The congress issued a variety of mostly Marxist–Leninist–styled manifestos and proclaimed the US armed forces in Europe to be the main enemy. It decided that the proper strategy was to kill individual soldiers in order to demoralize and weaken others.[55] The La Belle Discotheque bombing represented such a technique.

The La Belle Discotheque was a popular gathering spot for American servicemen in West Berlin. On April 5, 1986, a bomb exploded there, killing one American soldier and a Turkish woman and injuring 230 people, including some fifty Americans. The blast followed a March 28 call by Qaddafi for "all Arab people" to attack anything American, "be it an interest, goods, ship, plane or a person."[56] Indeed, it was not long before the attack was traced to Qaddafi in the form of the "smoking gun" that had thus far eluded American intelligence. Hours before the detonation, the East Berlin Libyan People's Bureau alerted Qaddafi, "We have something planned that will make you happy." After the attack, another message went out announcing, "An event occurred. You will be pleased with the result."[57]

President Reagan's objective was to limit Qaddafi's role in international terrorism rather than to eliminate him as the leader of Libya.[58] As Secretary of Defense Caspar Weinberger explained, "The purpose of our plan was to teach Qaddafi and others the lesson that the practice of terrorism would not be free of cost to themselves; that indeed they would pay a terrible price for practicing it."[59] Such an end is consistent with the "compellent" use of force that is designed "to cause an adversary to decide that further pursuit of its course of action would incur increasing costs incommensurate with any possible gain."[60]

The way the Reagan Administration sought to achieve such an end was "coercive diplomacy" which uses threats and limited, selective force "to induce the opponent to revise his calculations and agree to a mutually acceptable termination of the conflict."[61] In such an approach, the costs inflicted on the adversary are of a type and magnitude more geared to influence his decision than to physically impose one's will upon him. Thus the actions of the military units involved do not in and of themselves "attain the objective; goals are achieved through the effect of the force on the perceptions of the actor."[62] As such, Alexander George argues coercive diplomacy is "not a military strategy at all but rather a political strategy."[63]

With the end of persuading Qadaffi to reduce his terrorist activity and the way of coercive diplomacy determined, President Reagan and his advisors then had to consider the means they had available to employ. In the rational actor model, Russell Bova explains, "Key foreign policy decision-makers will collectively and systematically canvass the range of alternative policy choices from which they might choose and, ultimately, settle on the choice that will most effectively and efficiently promote national interests."[64] This responsibility fell to Vice Admiral Frank Kelso, commander of the US Navy's Sixth Fleet.

Kelso and his staff made use of the "military decision making process" (MDMP), a methodology that is consistent with the actor model and analytical decision-making. The MDMP is a way of "comparing courses of action against criteria of success and each other, selecting the optimum course of action."[65] The advantages of such an approach are as follows:

• It is methodical and allows the breakdown of tasks into recognizable elements.
• It ensures decision-makers consider, analyze, and evaluate all relevant factors.
• It provides a methodology when the decision requires great computational effort.

President Reagan being briefed by the National Security Council staff on the Libya air strike in the White House Situation Room on April 15, 1986. From left to right are President Reagan, Secretary of State George Shultz, CIA Director William Casey, White House Chief of Staff Don Regan, and General Charles Gabriel (Ronald Reagan Library).

- It provides a good context for decisions, especially for explanations.
- It helps resolve conflicts among courses of action.
- It gives inexperienced personnel a methodology to replace their lack of experience.[66]

As part of their MDMP, Kelso and his staff developed three courses of action for what would become Operation El Dorado Canyon. The first was to use special operations forces such as Navy SEALs or Army Special Forces to conduct raids utilizing helicopters, small surface boats, or scuba infiltration. The second was to use BGM–109C Tomahawk missiles. The third was to use sea- and land-based airpower.[67]

During the MDMP, the staff uses evaluation criteria to analyze and evaluate the advantages and disadvantages of each course of action (COA). The courses of action are then compared with each other to determine which has the highest probability of success against the enemy's most likely course of action and the enemy's most dangerous course of action. The most common technique to make this comparison is the decision matrix. The staff assigns numerical values for each of the evaluation criteria with the lowest number representing the best option. The scores for each criterion are then added together and the course of action with the lowest score is best.[68]

Criteria	COA 1	COA2	COA3
Simplicity	1	2	3
Ease of logistical support	1	3	2

(Criteria)	(COA 1)	(COA2)	(COA3)
Need for intelligence	3	2	1
Degree of collateral damage	2	3	1
Command and control	2	3	1
Total	9	13	8

AN EXAMPLE OF A DECISION MATRIX. IN THIS CASE, COA 3 HAS THE LOWEST SCORE AND THEREFORE IS THE BEST COA BASED ON THE EVALUATION CRITERIA SELECTED.

Kelso's staff considered such evaluation criteria as weapons accuracy, weight of payload, proximity of targets to civilian population centers, time over target, and survivability of attacking forces. The special operations option had the disadvantages of requiring lengthy planning and complex execution and the possibility of friendly casualties. The Tomahawk option was problematic because few missiles had been programmed for and were available for conventional missions in Libya. Moreover, if a missile was shot down and captured, the Libyans would likely turn over the sensitive technology to the Russians for exploitation. On the other hand, airpower represented the chance for a quick, precise, and punishing attack.[69] According to W. Hays Parks, "Tactical air offered the ability to place the greatest weight of ordnance on the targets in the least amount of time while minimizing collateral damage and providing the greatest opportunity for the survival of the entire force."[70]

Based on this rational analysis, the airpower option was selected as the one offering the optimal solution to the problem. The ensuing attack occurred on April 15, 1986, and the final strike force involved over one hundred Air Force and Navy strike and support aircraft from Europe and the Mediterranean. The targets were selected because of their association with terrorism and included training grounds, command and control headquarters, and airfields and aircraft.[71] Secretary Weinberger reported the results:

> The Sidi Bilal military complex was severely damaged. The Aziziyah barracks received substantial damage. The Tripoli International Airport was hit hard, and five IL–76/CANDID heavy transport aircraft on the apron were destroyed. The Benghazi barracks were hit and a warehouse in the complex, involved in MiG assembly, was destroyed. At Benina Airfield many planes were damaged or destroyed, including at least four MiGs; but most important, the Libyans were unable to launch planes from the airport during, or immediately after, the attack.[72]

One US Air Force plane and its two pilots were lost in the attack. Operation El Dorado Canyon was a clear "military victory."[73]

Beyond this tactical success, Secretary Weinberger notes that "the surest way to measure the success of an enterprise is to ask whether it achieved its objectives," which, in this case, "was to end Qaddafi's belief that he could use terrorism without cost."[74] Within the context of coercive diplomacy, Operation El Dorado Canyon did have an immediate, if not permanent, effect. Weinberger reports Qaddafi, who was wounded in the strike, maintained a very low profile after the attack and "pretty well vanished into the desert for an extended period." "Thus," Weinberger concludes, "our goals were realized, and one source of the export of terrorism was stopped at least temporarily."[75] Bruce Jentleson agrees that for a time after the attack, Qaddafi "appeared extremely disoriented" and "for nearly two years his role in terrorism fell off quite substantially."[76] Brian Jenkins goes even further, assessing that Operation El Dorado Canyon "permanently altered the equation. Any nation contemplating terrorist action against the United States after

On November 29, 1990, members of the United Nations Security Council voted to use "all necessary means" to uphold its resolutions if Iraq did not withdraw from Kuwait by January 15, 1991. It was this mandate that would guide US decision-making as to when to terminate combat operations (UN photograph by Milton Grant).

April 15, 1986, had to take into account the possibility of American retaliation."[77] As time passed, the effects of the attack diminished, and Qaddafi "reared his head anew," but for the time being, the rational actor model had produced a course of action that met the immediate objective of coercive diplomacy.[78]

Example 3: Bush's Decision to Halt
Operation Desert Storm, 1991

On August 2, 1990, Iraqi dictator Saddam Hussein launched an invasion of Kuwait. In spite of United Nations Security Council Resolution 660's demand "that Iraq withdraw immediately and unconditionally all of its forces to the positions in which they were located on 1 August 1990," the Iraqi forces quickly swept aside Kuwait's meager defenses, gained control of the country's oil fields, and forced Kuwaiti emir Jaber al-Ahmed al-Sabah into exile.[79] On August 8, Iraq announced the annexation of Kuwait. That same day, the United States deployed the first soldiers of the lead brigade of the 82nd Airborne Division to Saudi Arabia. By August 24, the 82nd had more than 12,000 soldiers on the ground.[80] Continued Iraqi intransigence led to the issuance on November 29 of UN Security Council Resolution 678 which authorized member states "to use all necessary means to uphold and implement resolution 660 (1990) and all subsequent relevant resolutions and to restore international peace and security in the area" unless Iraq was in compliance with all previous resolutions by January 15, 1991.[81] Based on this resolution, thirty-six countries, not including Kuwait, dispatched forces to the Persian Gulf.

The US–led coalition amassed 2,614 aircraft, 1,990 of which were American, in the Persian Gulf area. On January 17, the allied air offensive began and quickly established air superiority over Iraq. Iraqi air defenses were neutralized within a few hours, and allied forces then targeted command and control facilities, communications networks, airfields, transportation infrastructure, supply centers, and ground forces. In thirty-eight days, the coalition flew more than 90,000 sorties and claimed to have destroyed thirty-nine percent of the Iraqi tanks, thirty-two percent of the armored personnel carriers, and forty-eight percent of the artillery in Kuwait and southern Iraq. They also reported destroying more than one-third of the Iraqi aircraft. Equally important was the destruction of key bridges across the Euphrates River, which partially isolated the battlefield in preparation for the ground war.[82]

Coalition commander General Norman Schwarzkopf then orchestrated a massive deception campaign that convinced the Iraqis that the allied attack would occur in eastern Kuwait. Under cover of allied air superiority, Schwarzkopf shifted about 270,000 troops to the west, where Iraqi defenses were weakest. When the ground attack began on February 24, US Marines and other coalition forces attacked directly into eastern Kuwait. To their left, the VII Corps advanced about one hundred kilometers into Iraq and then turned east into the rear and flank of the Iraqis. On the coalition's west flank, the XVIII Airborne Corps raced north toward the Euphrates River to cut off the potential Iraqi escape route. This "left hook" advanced virtually unopposed.[83]

The devastation inflicted by the coalition forces was extraordinary. American intelligence estimates would later conclude about eighty-five percent of the Iraqi tanks, fifty percent of the armored personnel carriers, and ninety percent of the artillery in southern Iraq or Kuwait were damaged or destroyed. More than 10,000 Iraqi prisoners were taken in the first twenty-four hours of the ground attack and more than 70,000 total were ultimately captured. It is difficult to determine the number of Iraqi casualties, but the Defense Intelligence Agency estimates Iraq suffered 100,000 soldiers killed and 300,000 wounded, and that some 150,000 men deserted. A mere thirty-eight coalition soldiers died in the ground battle and seventy-eight were wounded.[84]

Amid such a lopsided environment, it was tempting for coalition forces to press on into Iraq and topple Saddam's regime. Instead, President George Bush ordered the "cessation of offensive operations" go into effect on February 28 at 8:00 A.M. The rational actor model is designed to effectively and efficiently achieve an objective, and as Schwarzkopf colorfully describes, the coalition objective was to "kick the Iraqi military force out of Kuwait." He notes the UN Security Council had provided "no authority to invade Iraq for the purpose of capturing the entire country or its capital."[85] Secretary of Defense Dick Cheney adds, "We'd told our Arab allies, the Saudis in particular, that we'd bring enough forces to liberate Kuwait and that we'd leave when we were done.... In addition, neither the United Nations nor the US Congress had signed on for anything beyond the liberation of Kuwait."[86] Various commentators debated the decision after the war, especially in light of Saddam's continued affronts to US national interests, but the cessation of offensive operations once the stated objective had been achieved is certainly as the rational actor model would predict.

President Bush addressed this issue in his memoirs, writing:

I still do not regret my decision to end the war when we did. I do not believe in what I call "mission creep." Our mission, as mandated by the UN, was clear: end the aggression. We did that. We liberated Kuwait and destroyed Hussein's military machine so that he could no longer threaten his neighbors.[87]

Bush's explanation is consistent with the assumption of the rational actor model that "once a policy choice is made, it will be implemented more or less as decision-makers had intended."[88] To now invade Iraq, as Chairman of the Joint Chiefs of Staff General Colin Powell describes it, would have been to depart from "a clearly defined UN mission" and "move the goalposts."[89] This consistency of outcome between plan and execution in decisions based on the rational actor model stems from the unity of purpose between those who define policy and those who carry it out.[90] Consistent with this assumption, Secretary of Defense Cheney writes, "No one on President George H. W. Bush's national security team was arguing in 1991 that we should continue on to Baghdad to oust Saddam."[91] Likewise, General Schwarzkopf reports that "at the time the war ended there was not a single head of state, diplomat, Middle East expert, or military leader who, as far as I am aware of, advocated continuing the war and seizing Baghdad."[92]

Still some critics have argued that the ultimate goal should have been to topple Saddam rather than merely liberate Kuwait. Even President Bush confessed, "I was convinced, as were all our Arab friends and allies, that Hussein would be overthrown once the war ended. That did not and has still not happened. We underestimated his brutality and cruelty to his own people and the stranglehold he has on his country."[93] Nonetheless, the rational actor model remains silent in explaining if, for the sake of argument, the stated objective of the UN mandate was too narrow, faulty assumptions were made, or the decision to cease combat was flawed in the strategic long-run. This model "does not assume that the policy ultimately chosen will prove to be the wisest and best of the available choices." It begins by identifying and defining the problem, identifying goals, and gathering information. The decision that emerges flows from this beginning, and an attempt to analyze the decision out of that context is a bastardization of the model as it is described. Likewise to declare that a certain decision-maker used the rational actor model does not imply that the observer shares or endorses the decision-maker's goals, conclusions, or methods. It merely suggests that the decision resulted from "a 'rational' process defined as the systematic effort to link policy ends and means."[94] That is exactly what happened in the decision to halt Operation Desert Storm.

Utility of the Rational Actor Model

The rational actor model is attractive because it is parsimonious and seems to be simple common sense. With a few straightforward assumptions, it can explain a host of decisions about foreign policy and the use of force.[95] It is especially compatible with the realist paradigm which regards states as unitary, rational actors.[96]

In spite of all its appeal, the rational actor model simply does not always emerge as the model used by decision-makers in the "real world." Various factors including a lack of accurate information or time, misperception, and human frailties compounded by stress all war against such a deliberate, structured, and organized process. Even more fundamental, however, is the suggestion that decision-making is not a completely rational act to begin with. As David Krantz explains, "People do and should act as *problem solvers, not maximizers*, because they have many different and incommensurable ... goals to achieve."[97] Indeed, many psychological considerations manifest themselves in ways that interfere with rational thought and make other decision-making models necessary.[98] Prospect theory is one such explanation.

2

Prospect Theory

Prospect theory is a theory of decision-making under conditions of risk based on a psychological model developed by Daniel Kahneman and Amos Tversky.[1] As summarized by Rose McDermott, "prospect theory predicts that people tend to be cautious when they are in a good position (gains), and more likely to take risks when they are in a bad position (losses)."[2] In the language of prospect theory, these relative positions are known as "domains" and are based on the decision-maker's perceived location.[3] This location is relative to a "reference point" which is "generally the current steady state, or status quo, to which a person has become accustomed."[4] In trying to maintain or regain this reference point, individuals display a pronounced aversion to loss, and prospect theory predicts a "greater focus and attention on real or feared losses than on prospective or foregone gains."[5] Kahneman and Tversky depict this expectation with an S-shaped value function that suggests changes closer to the reference point or status quo seem to count more psychologically than those occurring farther away from the origin.[6]

One of prospect theory's contributions as both an explanatory and predictive theory is that it accounts for the importance of the situation in the analysis of decision making.[7] Unlike norm-based models and rational choice theory, prospect theory conceptualizes human decision-making in terms of reason-based choice.[8] It characterizes the situation as domain and categorizes it in terms of gains and losses.[9] It is thus consistent with the reality that "people make decisions according to how their brains process and understand information and not solely on the basis of the inherent utility that a certain option possesses for a decision maker."[10] As a result, prospect theory is useful in understanding General George Washington's risky decision to cross the Delaware River and attack Trenton, Major General George Meade's cautious decision not to aggressively pursue General Robert E. Lee after the Federal victory at Gettysburg, and President Bill Clinton's political decision to end the US mission in Somalia in spite of the operational blow struck to warlord Mohammed Farrah Aideed during the Battle of Mogadishu.

Example 1: Washington's Decision
to Attack Trenton, 1776

By most accounts, George Washington's experiences and predilections had made him a conservative general.[11] Rather than with the "Audacity, audacity, always audacity" of Napoleon, Douglas Southall Freeman believes Washington emerged from the French and Indian War with the contrasting slogan of "Patience, patience, endless patience."[12] Russell Weigley echoes that Washington's very watchword was "caution." He certainly displayed a preference for carefully weighing the odds and acting accordingly. Pursuant to that philosophy, Washington advised Benedict Arnold as Arnold contemplated an attack against the British in Newport, Rhode Island:

> You must be sensible that the most serious ill consequences may and would, probably, result from it in the case of failure, and prudence dictates, that it should be cautiously examined in all lights, before it is attempted. Unless your Strength and Circumstances be such, that you can reasonably promise yourself a moral certainty of succeeding, I would have you by all means to relinquish the undertaking, and confine yourself, in the main, to a defensive operation.[13]

However, after the British threw 32,000 troops and almost half of the Royal Navy against New York City in August of 1776, Washington was forced outside his comfort zone. At the Battle of Long Island, Major General William Howe turned Washington from his strong positions, inflicting almost thirty percent casualties on the Americans while the British suffered about two percent.[14] By the end of November, Washington, "with mere remnants of his army, was in full retreat across New Jersey" with the British in hot pursuit. Washington's army was melting away with entire militia companies abandoning the cause and even his Continentals deserting in dangerous numbers. When Washington finally escaped across the Delaware River into Pennsylvania in early December he could muster barely 2,000 men.[15] Not only had his army failed to stand against the British regulars, he faced an impending catastrophic loss of manpower with many enlistments scheduled to expire on December 31. Surveying his tattered army, many of which were dressed only in blankets and rags, Washington confided to his brother on December 18, "I think the game is pretty near up."[16]

Thomas Paine captured the urgency of the crisis in a pamphlet called *The American Crisis*, which first appeared in the *Pennsylvania Journal* on December 19, 1776. "These are the times," Paine wrote, "that try men's souls. The summer-soldier and the sunshine patriot, will, in this crisis, shrink from the service of their country; but he who stands it *now*, deserves the thanks of man and woman."[17] One of these stalwart few was Washington, who Paine likened to a man who was only at his "full advantage, but in difficulties and in action."[18] Paine was also confident the public would rise to the occasion. "Panics," he argued, "in some cases, have their uses: they produce as much good as hurt." It is during panic, Paine claimed that "the mind acquires a firmer habit than before" and grows firmer, and "things and men" are brought to light "which might have otherwise lain forever undiscovered."[19] To this end, David Fisher describes *The American Crisis* as not merely an exhortation but a program for action. In this "pivotal moment when great issues of the Revolution where hanging in the balance," Fisher credits Paine with helping America resolve to act. Not just Washington and his army, but the fledgling nation as a whole was in the domain of loss. Fisher says, "most of all, it was a moment of decision, when hard choices had to be made."[20] In such cases, prospect theory would predict those hard choices would involve an increased willingness to accept risk.

Indeed, the Continental Congress responded to this moment of crisis by acting with unprecedented forcefulness. On December 21, it expedited government matters by appointing Robert Morris, George Clymer, and George Walton as "a Committee of Congress with Powers to execute such Continental business as may be proper and necessary to be done in Philadelphia."[21] While decentralizing authority in one direction, Congress centralized it in another. On December 27, it voted to grant Washington expanded, albeit temporary, powers to rebuild the army. As Oliver Wolcott described the measure in a letter to John Adams, "the whole of the military department is put into [Washington's] hand for six months.... The Preservation of the Civil Liberties of the People, at the present Time, depends upon the full exertion of the Military Power."[22] Squarely in the domain of losses wrought by the "December Crisis," Congress was willing to take risks. As predicted by prospect theory, when "commanded by that stern General, Necessity, [the delegates] had been compelled to approve what, in any other circumstances, they would have shouted down."[23]

Still, any change that might be wrought by these bold decisions would take time, and the present situation remained precarious. On December 22, Adjutant General Colonel Joseph Reed wrote Washington, "We are all of opinion my dear General that something must be attempted to revive our expiring Credit give our Cause some Degree of Reputation & prevent a total Depreciation of the Continental Money which is coming on very fast." Reflecting prospect theory's predictions about one's willingness to accept risk in such a clear domain of loss, Reed added, "even a Failure cannot be more fatal than to remain in our present Situation." Reed saw no alternative but that "some Enterprize must be undertaken in our present Circumstances or we must give up the Cause."[24]

General Washington, who Richard Ketchum argues was "frequently at his best when it appeared that things could not possibly get worse," was already thinking along the same lines as Reed. As if operating in the domain of loss summoned some latent energy in Washington, Ketchum considers him an example of those "few perverse souls [that] are always capable of doing their best after the chips are gone, when by all that is logical they should call it quits and give in to the inevitable."[25] Under such conditions, prospect theory predicts a willingness to accept risk, and Washington responded by planning what Robert Doughty describes as a "desperate effort" to strike a surprise attack on the Hessian garrisons at Trenton and Bordertown on Christmas night.[26] This tempting target was made possible because Howe, finding himself comfortably in the domain of gains after his string of victories, had retired to winter quarters in New Jersey, New York City, and Newport rather than pressing the best opportunities for trapping and destroying American forces that the British would have during the entire war.[27]

By the last week of December, Washington had assembled a force totaling about 7,000 men. He replied to Reed on December 23 that such numbers were "less than I had conception of — but necessity, dire necessity, will, nay must, justify any [attempt]."[28] Washington planned to cross the Delaware River at McConkey's Ferry above Trenton with a force of 2,400 Continentals under his personal command. Once on the New Jersey side, the force would divide into two columns proceeding on different routes designed to converge at Trenton from opposite directions in the early morning hours of December 26. A second force of mainly militia under Colonel John Cadwalader was to cross below Bordertown and strike the Hessians there while a third force of militia under Brigadier General James Ewing was to cross directly opposite Trenton to block the Hessian escape route across Assunpink Creek.[29]

While neither Cadwalader nor Ewing was able to fulfill his part of the plan, Washington boldly pressed on across the icy river, and the two columns converged on Trenton at 8:00 A.M. After a fight of just an hour and a half, the Hessians surrendered. Forty were killed and 918 taken prisoners compared to just four dead and four wounded among the Americans. The only disappointment was that, because Ewing had been unable to block the route, some four hundred Hessians managed to escape to Bordertown.[30] Washington quickly assessed the new situation and, after briefly considering the possibilities of pressing his victory, again acted consistent with the predictions of the prospect theory from his new domain of gains. He loaded up his army and withdrew back across the Delaware, having decided, in the words of his biographer Freeman, that he "must not take the chance of losing much by seeking more."[31]

Washington's success was yet a relative one, and he remained in the overall domain of losses with the still looming enlistment expirations. In such a threatening situation, Washington "played his last card" and offered a special bounty of ten dollars to each man who agreed to remain with the Army six weeks after the passing of his December 31 service obligation.[32] By this bold measure, Washington was able to muster a force of 5,000 men and again crossed the Delaware on the night of December 30–31. Washington avoided confronting a force Lieutenant General Charles Cornwallis had assembled at Trenton and instead launched a surprise attack at Princeton on January 3. There Washington inflicted heavy casualties on two British regiments and then went into winter quarters around Morristown, New Jersey.[33]

David Bonk calls the twin victories that occurred between December 25, 1776, and January 3, 1777, "ten days that shocked the world."[34] In the process, they did much to restore the Americans to their desired reference point. Henry Knox, Washington's chief of artillery, wrote his wife "The enemy were within nineteen miles of Philadelphia, they are now sixty miles. We have driven them almost the whole of New Jersey."[35] Perhaps equally importantly, as Maurice Matloff explains, "Trenton and Princeton not only off-set the worst effects of the disastrous defeats in New York but also restored Washington's prestige as a commander with friend and foe alike."[36] However, once Washington regained this comfortable status quo, he, as prospect theory would predict, became much more risk averse than he had been at Trenton. In fact, Allan Millett and Peter Maslowski note that "after 1776 Washington assumed the strategic defensive and determined to win the war by not losing the Continental Army in battle, fighting only when conditions were extremely advantageous."[37] No longer in the domain of losses, Washington subjected the British to a relatively safe war of attrition, confident that "by keeping his army in the field, by ignoring British attempts to bring him into decisive action and by waiting for the opportune moment," he could achieve victory.[38]

Example 2: Meade's Decision Not to Pursue Lee After Gettysburg, 1863

By 1863, General Robert E. Lee had become a problem for which President Abraham Lincoln had yet to find a solution. Upon assuming command of the Army of Northern Virginia in 1862, Lee had thwarted Major General George McClellan's drive to Richmond during the Peninsula Campaign. On the strength of that victory, Lee then humiliated Major General John Pope at Second Manassas. Hoping to decisively turn the tide of the war, Lee marched into Maryland where McClellan battled him to a tactical draw but strategic victory at Antietam. The Federal respite was short lived when Lee next trounced Major General Ambrose Burnside at Fredericksburg. In the aftermath of this latest debacle, Lincoln replaced Burnside with Major General Joseph Hooker.

Hooker's boast that "may God have mercy on General Lee for I will have none" was shattered on May 1 when Lee launched an attack on the Army of the Potomac at Chancellorsville. With the able Lieutenant General Stonewall Jackson crashing into Hooker's unprotected right flank, Lee inflicted 17,000 casualties on an enemy over twice his size. The Confederate victory was a devastating physical and psychological blow to both Hooker and the Federal Army. Long-gone was "Fighting Joe's" characteristic self-assuredness. "For once I lost confidence in Hooker," he confided to Major General Abner Doubleday. With the Federal Army and its commander reeling, Lee began planning a second invasion of Northern territory to build on this momentum.

On June 3, 1863, Lee started moving his Army of Northern Virginia north from Fredericksburg. Still stinging from Chancellorsville, Hooker proceeded cautiously. Eventually he suggested to President Lincoln that, now that Lee was gone, the Federals could take Richmond. Lincoln responded, "I think Lee's army, and not Richmond, is your true objective point."[39]

As Hooker timidly shadowed Lee on a parallel route, President Lincoln tried to press the Federal commander to strike. Hoping to spur some action, Lincoln asked Hooker where Lee's army was and Hooker replied, "The advance is at fords of the Potomac and the rear at Culpepper Court House." That represented a distance of seventy miles and Lincoln, even without formal military schooling, saw an opportunity. "If the head of the animal is at the fords of the Potomac and the tail at Culpepper Court House, it must be very thin somewhere," Lincoln advised Hooker. "Why don't you strike it?" For Lincoln, any situation in which Lee retained his freedom of maneuver placed the Federal cause in the domain of loss.

Hooker belatedly developed a plan to strike the extended Confederate line of communication, and he ordered his XII Corps to move to Harpers Ferry, combine with other forces there, and attack Lee's rear. When Major General Henry Halleck, who Hooker had increasingly felt was interfering with the Army of the Potomac, countermanded this order, Hooker asked to be relieved. Lincoln had long since grown frustrated with Hooker and was eager to take this opportunity to replace him with little political repercussion. On June 27, Major General George Meade, former commander of the V Corps, was ordered to assume command of the Army of the Potomac. With some understanding of Meade's apprehension, Halleck noted, "considering the circumstances, no one ever received a more important command."[40]

It was not a position Meade had sought, and he was overwhelmed and intimidated by the responsibility. He described the order as "totally unexpected" and confessed "ignorance of the exact condition of the troops and position of the enemy."[41] "Why me? Why not [Major General John] Reynolds?" Meade complained. "I don't know the Army's position. I don't know its plans. I don't know if it has any plans."[42] He wrote his wife, lamenting, "You know how reluctant we both have been to see me placed in this position." Obviously strained by the massive responsibilities, he mused to his wife, "Oh, what I would give for one hour by your side to talk to you, to see my dear children and be quiet!"[43] Meade, however, resolved to do his duty, acknowledging the order, "As a soldier I obey it, and to the utmost of my ability will execute it."[44] Privately, he was less resolute, reportedly confiding to Colonel James Hardie, who delivered him the order, "Well,

In spite of his victory at Gettysburg, Meade and his army had been severely tested as evidenced by the photograph's caption: "A harvest of death" (Library of Congress, Prints & Photographs Division, LC-B8184–7964-A DLC).

I've been tried and condemned without a hearing and I suppose I shall have to go to the execution."[45]

Kahneman and Tversky identify two phases of the choice process: "an early phase of editing and a subsequent phase of evaluation." In the editing phase, the decision-maker conducts "a preliminary analysis of the offered prospects, which often yields a simpler representation of these prospects."[46] Meade's analysis placed him in a negative situation which almost assumed defeat and incrimination. He overestimated the size of Lee's force, concluding the Confederate had more than 100,000 men when he only had 75,000.[47] Edward Stackpole attributes this exaggeration of Lee's strength in part by claiming Meade was "apparently obsessed with the myth of Lee's invincibility."[48] Indeed, Meade understood he would have to face this daunting opponent in "a battle that will decide the fate of our country and our cause."[49] In the midst of such pessimism and high stakes, victory would be defined as merely escaping catastrophe. It was an inauspicious beginning for the fifth commander of the Army of the Potomac, and in such circumstances, Meade was predisposed to follow the prospect theory's prediction of aversion to loss.

Gettysburg was a three day fight. The first day was an unplanned meeting engagement which boiled down to a contest to see which side could bring up reinforcements the fastest. July 1 ended in a Confederate victory, but the Federals were able to remain in control of the key terrain of Cemetery Hill.

On July 2, Lee ordered Lieutenant General James Longstreet to make the main attack against Meade's south flank. In the nick of time, Federal forces arrived at Little Round Top and held the position. The second day ended for the most part as a draw.

On the third day, Lee ordered a massive assault on the Federal center. The attack, including the famous Pickett's Charge, was repulsed. Upon hearing the news, all an exhausted and relieved Meade could muster was "Thank God."[50] Nonetheless, the Federals could claim victory on the third day and for the overall battle.

Now the question became what to do with the victory. Both sides had been badly hurt with Lee suffering 28,000 losses and Meade 23,000. The two armies lay facing each other "like spent lions nursing their wounds."[51] Even the triumphant Meade now commanded an army that was battered by its victory. His men were tired from days of marching and fighting. Dead and wounded soldiers and damaged and discarded equipment filled the battlefield and had to be recovered. Many veteran soldiers, as well as key leaders such as John Reynolds and Winfield Scott Hancock, were killed or wounded.[52]

Meade, too, was personally exhausted. He was, in the words of historian Bruce Catton, "on the road with his troops, an infinitely weary man with dust on his uniform and his gray beard, feeling responsibility as a paralyzing weight."[53] After the battle, a newspaper reporter found Meade "stooping and weary," and Freeman Cleaves, Meade's biographer, describes him as "a picture of sorry discomfort."[54] True the Federal cause had been saved, but Meade no doubt felt "even though only as one escaping through the flames."[55]

Meade was in no mood to exploit his victory, a view shared by many of his subordinates and other observers. Chief engineer Major General Gouverneur Warren spoke for many when he described the feeling "that we had quite saved the country for the time and that we had done enough; that we might jeopardize all that we had done by trying to do too much."[56] Former Army of the Potomac commander Major General George McClellan congratulated Meade, telling him, "You have done all that could be done."[57] Local resident and subsequent historian of the Gettysburg Campaign, Jacob Hoke explains, "General Meade, it should be remembered, had been in command of the army but six days. The responsibility which was thrust upon him was great indeed. A false step at the juncture under consideration would have resulted most disastrously. He had to decide, not simply for the time, nor for the army under his command, but for the whole country, for the Government, and for all time to come."[58]

Indeed, Meade's brief tenure in command had moved him from the domain of loss to the domain of gain. As Catton explains, Meade "had been one of the few men who could have lost the war irretrievably in one day, and he had managed to avoid the mistakes that would have lost it."[59] As prospect theory would predict, Meade would now be loss averse. According to Hoke, such a situation left Meade with an "inclination to be on the safe side" and "resist putting in jeopardy all he had already gained."[60] Catton agrees, explaining that Meade "would continue to avoid mistakes, even if he had to miss opportunity ... Meade could see all the things that might go wrong."[61]

Lincoln did not see these challenges; only the opportunity. As president of a divided nation, he remained in an overall domain of loss and felt Meade's work was unfinished. Thus, it was not time to celebrate yet, and Lincoln announced Meade's success to the nation in restrained tones, withholding from Meade the thanks and praise Lincoln had lavished on Major General Ulysses Grant after his victory at Vicksburg.[62] In place of congratulations, Meade was told, "The opportunity to attack [Lee's] divided forces should not be lost. The President is urgent and anxious that your army should move against [Lee] by forced marches."[63]

Instead, Meade acted as if his mission was accomplished. Some observers comment that he appeared to have "escorted" Lee out of the North rather than pursuing him bent on his destruction. Watching Meade's halfhearted effort, Lincoln complained, "I'll be hanged if I could think of anything but an old woman trying to shoo her geese across a creek."[64] Lincoln was beside himself, writing Meade a letter he refrained from sending that said, "I do not believe you appreciate the magnitude of the misfortune involved in Lee's escape.... He was within your easy grasp, and to have closed upon him would, in connection with our other late successes, have ended the war. As it is, the war will be prolonged indefinitely."[65] As Stackpole explains, "Some generals play it safe, forgetting perhaps that war is far from being a safe enterprise. Others take calculated risks when large results are possible. The great Captains have been those who audaciously and aggressively discounted the odds, whether actual or imagined, and by their boldness won important victories.... Meade was not [that kind of general]."[66]

Stackpole's assessment notwithstanding, perhaps a more nuanced explanation of Meade's choice in this particular situation would be that Lincoln and Meade had different reference points. In his congratulatory order to his army, Meade had written, "Our task is not yet accomplished and the commanding general looks ... for greater efforts to drive from our soil every vestige of the presence of the invader."[67] On the other hand, Lincoln complained to Halleck, "I did not like the phrase ... 'Drive the invaders from our soil.'" Lincoln considered the entire country to be United States soil. He wanted Lee crushed and was frustrated that Meade's post-battle actions "appear to me to be connected with a purpose to ... get the enemy across the river again without a further collision, and they do not appear connected with a purpose to prevent his crossing and to destroy him."[68] Lincoln wanted nothing short of total victory. He wrote Halleck, "Now if General Meade can complete his work, so gloriously prosecuted thus far, by the literal of substantial destruction of Lee's army, the rebellion will be over."[69]

Because prospect theory recognizes that the same situation can be understood as either an opportunity to gain or a chance to avoid loss, it allows the observer to consider an event from the participant's perspective.[70] As president, Lincoln had a much grander view of the situation than Meade had as operational commander. Lincoln's desired status quo was the restoration of the Union. Meade's desired status quo was merely the battlefield equilibrium that existed before Lee's invasion of the North. These different reference points allowed each man to interpret the same event differently. For Meade, the repulse of Lee at Gettysburg represented a transition from the domain of loss to the domain of gain. For Lincoln, this operational outcome was a welcome improvement, but from a strategic perspective, he still viewed himself in the domain of loss. He complained to his private secretary John Hay, "Our Army held the war in the hollow of their hand & they would not close it." In imagery that clearly depicts an incomplete gain, Lincoln added: "We had gone through all the labor of tilling & planting an enormous crop & when it was ripe we did not harvest it."[71]

To be fair to Meade, any analysis of his actions must be done not from the comfortable view of the detached observer, but from the much more tenuous position that was Meade's perspective. When Lincoln pressed Meade to move against Lee "by forced marches," Meade must have been incredulous. From his point of view, "up to now there had been nothing but forced marches both before and after Gettysburg."[72] He tersely wrote Halleck, "My Army is and has been making forced marches short of rations and barefooted.... Our corps marched yesterday and last night over 30 miles. I take occasion to repeat that I will use my utmost efforts to push forward this Army."[73] While Meade thought he was doing everything humanly possible in the situation he found himself in on the battlefield, Halleck explained he was only trying to relay "opinions formed from information received here" in Washington.[74] Indeed, statements by Lincoln clearly show an oversimplification of the matter. He boasted to his son Robert, "If I had gone up there I could have whipped them myself."[75] To Major General Oliver Howard, Lincoln wrote, "I was deeply mortified by the escape of Lee across the Potomac, because the substantial destruction was perfectly easy."[76] From his perspective, Meade was likely to agree with Carl von Clausewitz that "everything in war is simple, but the simplest thing is difficult."[77]

Prospect theory is instructive in such cases because it is "concerned with the importance and impact of the environment on the person."[78] Under such circumstances, it is easy to understand why Meade succumbed to such an attitude of loss aversion, or, as he described it in a letter to his wife, to refuse to do "impractical things."[79] He had assumed command under extremely traumatic circumstances and turned back the great Robert E. Lee from Northern soil. That, from the reference point Meade had ascribed for himself, was enough.

Example 3: Clinton's Decision to Withdraw from Somalia, 1993

After anarchy, drought, civil war, and banditry had reduced Somalia to a virtual wasteland, in December 1992 the United Nations Security Council approved Resolution 794, which established Unified Task Force (UNITAF), a large, US-led peace enforcement operation known as Operation Restore Hope. The United Task Force made great progress, and humanitarian agencies soon declared an end to the food emergency. In light of these improvements, US forces began withdrawing in mid–February, 1993, and on May 4, United Nations Operation in Somalia (UNOSOM) II, armed with a much broader mandate based on a new United Nations Security Council Resolution 814, took over operations from UNITAF. United Nations Operation in Somalia II's mission was to "conduct military operations to consolidate, expand, and maintain a secure environment for the advancement of humanitarian aid, economic assistance, and political reconciliation."[80] This more aggressive posture soon brought US forces into increasing confrontation with Somali warlord Mohammed Farrah Aideed, who then perceived himself in the domain of loss.

On October 3, US special operations forces launched Operation Gothic Serpent, and Army rangers and Delta commandos conducted a daylight raid on a suspected location of Aideed and his lieutenants at the Olympic Hotel in Mogadishu. The Americans captured twenty of Aideed's men, but the mission quickly unraveled when Somalis shot down three US helicopters. The Americans soon became surrounded by thousands of Somalis, and the relief column was ambushed on its way to rescue the beleaguered soldiers. It was more than nine hours before help, primarily from the 10th Mountain Division, finally arrived.

Eighteen precious American lives were lost in the fighting, but by all accounts the battle was an overwhelming tactical victory. Mark Bowden, author of *Black Hawk Down*, reports as many as five hundred Somalis had been killed and over a thousand wounded.[81] Ambassador Robert Oakley, who arrived in Mogadishu on October 9 as President Bill Clinton's special envoy, offers his "own personal estimate ... that there must have been 1,500 to 2,000 Somalis killed and wounded that day."[82] More than mere numbers, many of the casualties came from families aligned with Aideed, and the losses took a toll on his position within the tangled web of Somali clan-based political affiliations.[83] Among those captured were several of Aideed's key lieutenants, including Mohammed Awale, a member of Aideed's central committee.[84] There were reports that some of Aideed's strongest clan allies had fled Mogadishu in fear of an American counterattack and others were willing to abandon Aideed in exchange for peace.[85]

Even before the October battle, some analysts had found indications "that UNO-SOM II operations in Mogadishu and in the rest of Somalia were seen by most Somalis as continuing signs that Aideed would ultimately lose his struggle with UNOSOM II. As a result, support for him within his clan was eroding." According to this view (and as prospect theory would suggest), Aideed's escalation of the violence was actually a sign of weakness and a desperate attempt to split the international coalition and weaken its resolve.[86] As a result of these devastating new casualties on top of his already tenuous grip on power, Aideed, according to one US general, was "was on the ropes."[87]

Believing Aideed "had been struck a mortal blow," many military leaders thought "it wouldn't take much more to finish the job."[88] In the immediate aftermath of the battle, UNOSOM II Force Commander Lieutenant General Cevik Bir prepared a memorandum for Admiral Jonathan Howe, the Special Representative to the UN Secretary General for Somalia, reflecting the opinion of several senior USOSOM II personnel that an unparalleled opportunity existed to exploit the military situation created by Aideed's vulnerability.[89] Former UNITAF J3 Major General Anthony Zinni also sensed the momentum was now clearly with the UN forces. Zinni had returned with Oakley to Somalia, and before meeting with Aideed, Zinni encountered some "really hard core militia fighters that were involved in the battle." According to Zinni, they "were extremely shook up." He reports a couple of them approached him and said, "No more, we have to stop the shooting." In Zinni's mind, "the battle that day really took its toll," and "the incident probably shook up Aideed and his militia to an extreme amount." Citing "the closeness of maybe grabbing some very key people [and] the number of casualties that they suffered in terms of militia," Zinni found the veteran fighters appearing "really down and worried." He found the same when he met with Aideed, reporting, "I'd never seen him like that before, I think that was probably the most frightened I saw him."[90]

If the military perceived itself to be operationally in the domain of gains after the Battle of Mogadishu, President Clinton and the US public certainly considered themselves to be in the strategic domain of losses. The October 18, 1993, covers of *U. S. News & World Report*, *Time* and *Newsweek* all featured pictures of captured Black Hawk pilot Chief Warrant Officer Michael Durant. *Time's* cover asked, "What in the world are we doing?" *U. S. News & World Report* blared "Somalia: What Went Wrong." *Newsweek* proclaimed, "Trapped in Somalia." Prospect theory predicts that decision-makers, even in deteriorating situations, will take risks "in the hope of recovering past losses, turning the tide of battle, or recouping sunk costs."[91] McDermott notes, "Particularly when there are many battle deaths involved, it can be politically difficult for leaders to cut and run. If they do, questions why these soldiers died in the first place can haunt them into early and undesired retirement."[92] Indeed, a day after the Battle of Mogadishu, President Clinton told aides, "I'm just not going to have those kids killed for nothing."[93] Even when he addressed the nation on October 7, Clinton acknowledged General Colin Powell's admonishment: "Because things get difficult, you don't cut and run. You work the problem and try to find a correct solution." Nonetheless, Clinton went on to announce "an orderly withdrawal" that would remove all but "a few hundred support personnel in non-combat roles" from Somalia by March 31, 1994.[94] Although Clinton resisted the temptation to take risks in the domain of loss, his decision-making can still be explained in terms of the prospect theory.

First of all, domains are relative to a reference point that is subject to change. When measured against UNOSOM II's expanded nation-building mission and mandate as prescribed by United Nations Security Council Resolution 814, recent developments in Somalia clearly put Clinton and the US in the domain of losses. Clinton could not change the facts on the ground, but he could change the yardstick against which they were measured. He created this new relationship by declaring, "It is not our job to rebuild Somalia's society."[95] By establishing a new reference point, Clinton instantly improved his position.[96]

A second factor is that the US presence in Somalia was a reality Clinton had inherited from his predecessor President George Bush; a fact Clinton tactfully reminded the American public in his October 7 speech.[97] Several studies have shown that "individuals who are not responsible for an original policy that failed may be more likely to be willing to stop the policy, as opposed to putting further resources into it."[98] This phenomenon allowed President Clinton to frame the situation not solely in terms of gains and losses in Somalia, but in terms of his larger presidency. By this time, Clinton was grappling with what many critics considered underdeveloped policies regarding not just Somalia, but Bosnia and Haiti as well. A common theme was that Clinton had "drifted away from his campaign promise to put US economic interests first in the conduct of foreign affairs.... Instead of a hard-headed emphasis on economics, Clinton has sometimes become preoccupied with his own naive visions of a pro-active, humanitarian presidency."[99] Extricating himself from Somalia presented Clinton with the opportunity to move closer to his desired reference point of a domestic-focused presidency.

Finally, Clinton was able to generate domestic political support for a withdrawal by justifying the action in terms of loss avoidance.[100] In evaluating decisions like Clinton's in light of prospect theory, Robert Jervis reminds that "we may have to shift our attention from the decision-makers to the beliefs and values of those who can reward and punish them."[101] As a peripheral interest, Somalia had never evoked a deep US commitment, and domestic support for the operation was weak. David Rieff explains the public perception of its reference point:

> The American public came to think of the hunt for Aideed, even though they knew it was being carried out by US Army Rangers, not as war but as police work. Casualties in war are understood to be inevitable. Soldiers are not only supposed to be ready to kill, they are supposed to be able to die. But casualties in police work are a different matter entirely. There, it is only criminals who are supposed to get hurt or, if necessary, killed, not the cops. Again, the fundamental problem has not been some peculiar American aversion to military casualties. Rather, there has been an essential mistake in the way such operations are presented to the public, and, perhaps,

even in the way they are conceived of by policymakers. Under the circumstances, it should hardly be surprising that public pressure on Congress and the president to withdraw US troops predictably arises at the first moment an operation cannot be presented in simple moral terms, or when casualties or even the costs start to mount.[102]

As a result of this mentality, after the October 3 casualties, one survey reported an overwhelming eighty-nine percent of respondents felt an "important goal of the US in Somalia" should be "bringing US troops home as soon as possible." Less than half that number now felt it was important for the US to establish "a stable government in Somalia."[103] Thus the American public and President Clinton were willing to accept "a small but sure loss [largely a blow to national and personal prestige] in order to eliminate the possibly of a larger catastrophe [more casualties and a Vietnam-like quagmire]."[104]

After the Battle of Mogadishu, Mark Bowden concludes Aideed "didn't know it yet, but his clan had scored a major victory."[105] American military leaders eager to press their tactical advantage found themselves powerless against a wave of loss aversion and a recalculated reference point by which President Clinton changed the US perception of its role in Somalia. Furthermore, as the principal decision-maker in the matter, President Clinton sought a return to the domestically-focused status quo upon which he had campaigned. Shedding himself of an inherited overseas situation helped move him in that direction. The decision to withdrawal from Somalia shows the flexibility of the prospect theory to accommodate political and psychological factors into decision-making.[106]

Utility of the Prospect Theory

Although applications of prospect theory to international relations have grown, the theory enjoys much broader acceptance in the fields of economics and psychology.[107] One explanation for the tepid response of political scientists is the methodological argument that "studying choices between gambles in the lab is different from studying complex political decisions in the field, where precise measurement of domain, risk, and the influence of other factors is impossible."[108] Many skeptics are dissuaded by "the difficulty of specifying in a nontautological fashion decision-makers' reference points and their calculations of gains and losses."[109] Moshe Levy and Hiam Levy go as far as to reject Kahneman and Tversky's S-shaped function, arguing instead that it "is actually due to the well-known certainty effect and does not represent investors' preferences in a realistic setting of mixed bets."[110] James Goldgeier and Philip Tetlock posit that political scientists may have some cultural bias against prospect theory, noting "it seems likely that the resistance is to the discipline of psychology rather than to problems unique to prospect theory."[111]

Champions of the theory are equally adamant in defending their cause. Jonathan Mercer argues, "Prospect theory is not a fad, a curiosity, or a way to capture idiosyncratic behavior. It is the most influential behavioral theory of choice in the social sciences." Like any theory, prospect theory has its problems and limitations, but Mercer insists its practioneers are "methodologically sophisticated."[112] Arguing their findings "actually support prospect theory," Peter Wakker criticizes Levy and Levy for "overlooking the crucial role of probability weighting in prospect theory."[113] Numerous observers refer to the intuitive attractiveness of prospect theory, noting that "perhaps the main reason for thinking the effect is real is that the tendency for people to be risk acceptant for losses resonates with personal experience."[114]

Rose McDermott is perhaps the most well-known champion of prospect theory among political scientists due to her *Risk-Taking in International Politics: Prospect Theory in American Foreign Policy* (University of Michigan Press, 1998) and her well-reasoned application of the theory to President Jimmy Carter's decision to launch the ill-fated Iranian hostage rescue mission. In spite of her attachment to the theory, McDermott provides a well-balanced assessment of the subject in "Prospect Theory in Political Science: Gains and Losses from the First Decade" (*Political Psychology*, April 2004). She identifies a lack of development of a theory of framing, lack of applicability to group behavior, and lack of attention to issues surrounding the importance of emotion as limitations, but argues these may be overcome with future research.[115] Still, she insists that "the utility of applying prospect theory to case examples in political science has been established," and it is time to move on to further work that extends prospect theory's interaction with other theoretical perspectives.[116] As no single theory can by itself reliably explain the wide range of decision-making processes, it is hard to argue against such an inclusive approach. Indeed, poliheuristic theory endeavors to explain decision-making as both a cognitive and rational process.

3

Poliheuristic Theory

Based on pioneering work begun by Alexander Mintz in 1993, poliheuristic theory holds that decisions are made as a result of a two-stage process. Mintiz and colleague Karl DeRouen describe their theory as "dimension based, non-compensatory, nonholoistic, satisficing, and order sensitive."[1] First, the decision-maker simplifies the problem by using cognitive short-cuts or heuristics. This quick first cut eliminates any option that involves any unacceptable political costs. Indeed, poliheuristic theory sees domestic politics as "the essence of decision."[2] Then, the decision-maker evaluates the remaining alternatives using a detailed, rational cost-benefit analysis. One of the advantages of such an approach is that it saves time and makes the decision-making process more manageable. However, because avoiding political loss is often the driving goal for state leaders, a complete assessment of costs and benefits may not be the prime motivation in the first stage of the decision. As a consequence, the best option may be discarded before being exposed to the analytical calculations of the second stage. The result may often be suboptimal "satisficing" or settling for an outcome that minimally satisfies a more limited set of objectives rather than maximizing the national interest.[3]

A common heuristic decision-makers apply during the first stage is the non-compensatory principle.[4] This principle posits that "a low score on one dimension cannot be compensated for by a high score on another dimension."[5] When applied to a foreign policy decision, this means that the political dimension is the paramount attribute and that "in a choice situation, if a certain alternative is unacceptable ... politically, then a high score on another dimension ... *cannot* compensate/counteract for it, and hence the alternative is eliminated."[6] Especially in highly charged foreign policy crises, decision-makers can be expected to resort increasingly to non-compensatory strategies in an effort to simplify the decision-making process and avoid political losses.[7] The trade-off is that not all information is reviewed and all alternatives are not considered on all dimensions before an acceptable solution is found.[8]

Poliheuristic theory bridges the gap between cognitive and rational theories of decision-making. The first stage, with its use of decision heuristics, corresponds to cognitive psychology, while the second stage, by its use of an analytic process of surviving alternatives, corresponds to rational choice theory.[9] President Dwight Eisenhower's refusal to intervene in the Hungarian revolt, President John Kennedy's authorization of the Bay of Pigs operation against Cuba, and President Lyndon Johnson's decision not to escalate the war in Vietnam after the Tet Offensive all contain elements consistent with the poliheuristic theory.

Example 1: Eisenhower's Decision Not to Intervene in Hungary, 1956

The term poliheuristic can be broken down into "the roots poly (many) and heuristic (shortcuts), which alludes to the cognitive mechanisms used by decision makers to simplify complex foreign policy decisions."[10] A common heuristic included in this number is the concept of "elimination by aspects" first articulated by Amos Tversky in 1972. In this model, the decision-maker chooses an aspect or attribute, eliminates unacceptable alternatives on the basis of that attribute, and continues the process until the remaining alternatives do not share any common aspects. Elimination by aspects is a non-compensatory approach in that the process is simplified by sequentially eliminating alternatives that do not meet a certain threshold in one, or a few, criteria as opposed to the compensatory process that requires a comparison of alternatives across all dimensions.[11] Elements of elimination by aspects theory are present in the Eisenhower Administration's response to the Hungarian revolt in 1956.

With the death of dictator Joseph Stalin in March 1953, Soviet satellites in Eastern Europe began pushing the limits of "de-Stalinization." In June 1956, Polish workers rioted in Poznan, and Polish troops responded with a crackdown that left fifty-three dead and over three hundred wounded. Then on October 23, peaceful student protests in Hungary turned into a violent revolt that quickly spread. At the time of the outbreak, the Soviets had two mechanized divisions in Hungary. Although the Hungarian army and regular police either joined the rebels of refused to fight against them, the Allamvedelmi Hatosag (AVH), Hungary's secret police, sided with the Soviets, and there were soon reports of massive losses of life. For example, on October 25, Soviet troops fired on a crowd outside the Parliament Building in Budapest, killing at least three hundred.[12] President Dwight Eisenhower lamented, "Poor fellows, poor fellows. I think about them all the time. I wish there were some way of helping them."[13] On October 26, he ordered a comprehensive study of the situation in order to determine possible courses of action.[14]

There was no shortage of ideas. Deputy Undersecretary of the Department of State Robert Murphy claims "everyone in the United States wanted to help the brave Hungarians," and the State Department soon was inundated with proposals ranging from military action to humanitarian assistance.[15] The decision-making process quickly assumed characteristics of the elimination by aspects model.

The political rhetoric at the outbreak of the crisis would suggest a decisive American response. Many in the Eisenhower Administration had outspokenly criticized President Harry Truman's policy of containment as not going far enough to confront communism. Instead, advisors like President Eisenhower's Secretary of State John Foster Dulles advocated an effort to "roll back" communism in Eastern Europe. The Republican platform of 1952 had promised a foreign policy that would "mark the end of the negative, futile and immoral policy of 'containment' which abandons countless human beings to a despotism of godless terrorism which in turn enables the rulers to forge the captives into a weapon for our destruction."[16]

Eisenhower had also made it clear that he disagreed with Truman's attempts to balance nuclear and conventional forces. Eisenhower believed he could fulfill his campaign promise to reduce federal spending by placing greater reliance on nuclear weapons. The result was a "New Look" in national security policy that declared, "In the event of hostilities, the United States will consider nuclear weapons to be as available for use as other munitions." Dulles took this strategy still further in January 1954 when he called for "massive retaliation" in response to Soviet aggression.[17]

In an era in which the US still enjoyed a strategic nuclear superiority against the Soviets, such aggressive action in the case of Hungary was at least a theoretical possibility.[18] Robert Amory, the CIA's deputy director for intelligence, was among those that proposed the use of tactical nuclear weapons. Give the Soviets an ultimatum, Amory urged, to "keep their hands off Hungary or we would not be responsible for whatever happened next." Amory argued that a "very small surgical strike at a few marshaling yards in those narrow valleys would seriously disrupt the Soviets' ability to move troops into Hungary." Yet even Amory conceded such an action would have immediately "fired up Soviet preparations for general war."[19] Such an outcome had serious ramifications, and Undersecretary Murphy reports all proposals were first evaluated based on the aspect of whether the action "would inevitably lead to a direct military attack by American forces against Russian forces."[20] After some discussion of Amory's proposal, such aggressive actions were eliminated, and the pool of American alternatives was reduced to only "action short of war."[21]

Two major considerations mitigated against a bellicose American response. First of all was the fact that the Soviets possessed conventional forces superiority in Hungary and the immediate area. Hungary did not border any NATO country, and a US intervention would require a violation of Austria's neutrality and even then leave the force vulnerable to a Soviet flank attack through Czechoslovakia. Furthermore, Europe had clearly settled into a US sphere of influence in the west and a Soviet sphere of influence in the east. Any intervention by one power into the realm of the other would likely precipitate an unacceptably wider war.[22] Even the Yugoslav ambassador to the US appealed to Murphy to avoid any action that would spread the war outside Hungary.[23] By the end of October, this fear took on new proportions when France and Great Britain used the clash between Egypt and Israel to reclaim the Suez Canal, heightening tensions between the US and the USSR.[24]

Aside from these fundamental considerations of power, there was also the American belief that any action that led to the downfall of Soviet leader Nikita Khrushchev would not be in the interests of the US. Khrushchev's program of de-Stalinization signaled the beginning of a détente with the West, and the Soviet-American relationship was transitioning from its previous volatility to a more stable, long-term competition. In the spring of 1956, Khrushchev had even announced plans for a reduction of more than a million men in the Soviet armed forces. Such measures clearly marked him as the most "dovish" of the Soviet leaders, especially when compared to hardliner alternatives such as Vyacheslav Molotov and Lazar Kaganovich.[25]

An attempt at some middle ground that would support the Hungarians without directly confronting the Soviets was to airdrop supplies to the rebels. However, this alternative was eliminated because flying over communist-controlled East Germany, Czechoslovakia, or Yugoslavia was too dangerous and provocative, and neutral Austria made it clear it would resist any overflights.[26]

Airdropping supplies also constituted a part of a covert plan that was considered once direct action had been eliminated. Deputy CIA Director Frank Wisner championed a proposal to insert émigré battalions into Hungary to aid the insurgency, but this action was criticized because of doubts of their ability to stand up to Soviet tanks. The small force would have to be appropriately armed, and the CIA conceded "there were no weapons handy enough to commandeer hurriedly, we knew too little, we had absolutely no picture at all of who needed weapons, when, what kind, where."[27] Logistics, intelligence, and readiness all conspired against covert action, and it was eliminated as an option.

The US next turned to diplomatic efforts in the United Nations. A Security Council Resolution affirming Hungarian self-determination and calling upon the USSR to cease its intervention was predictably vetoed by the Soviets on October 28. An exasperated Murphy reports, "We pushed as hard as we could in the UN forum, but we failed in our essential objective — release of the Hungarian people from bondage."[28]

In addition to Soviet intransigence, diplomatic efforts were also hamstrung by Hungarian Premier Imre Nagy's initial handling of the crisis. In a series of radio broadcasts on October 24, Nagy called for order and established a military tribunal which was authorized to summarily sentence anyone involved in resistance activities. He also claimed that the Hungarian government had asked the Soviet troops to enter Budapest.[29] Based on such statements, on October 25, American Ambassador to the Soviet Union Charles Bohen notified the State Department, "In view of [the] appeal by Nagy for Soviet troops there is not on the surface at least, any open differences between [the] Hungarian and Soviet Governments."[30]

One of the dangers of the elimination by aspects model is that once eliminated, options are seldom reconsidered. Thus once the American government concluded Nagy was "in complete collaboration with Moscow ... it never abandoned this view [even] as events changed."[31] This change occurred on November 1 when Nagy announced that Hungary was pulling out of the Warsaw Pact, declared its neutrality, and appealed to the United Nations for immediate assistance. Although this new development represented a "fundamental change in the nature of the crisis, American policy did not alter." It is true that the obstacles that impeded US military action still remained, but new diplomatic options such as recognizing Hungary's neutrality which at least would have made Soviet aggression more costly in the eyes of the world were now possible. Apparently, however, such a course was never seriously considered. Instead, on November 2 the US authorized $20 million in food aid to Hungary — "food [that] was little help to a government struggling for its very existence under the shadow of Soviet tanks."[32] The diplomatic effort climaxed on November 4 when the UN General Assembly passed a largely symbolic resolution "condemning the use of Soviet military forces to suppress the efforts of the Hungarian people to reassert their rights."[33] By then, the Hungarian revolt had been crushed, and the vote merely put on record the fact that the overwhelming majority of UN members protested the Soviet action. "Coming ex post fact," one scholar notes, "it was like whistling in the wind."[34]

"In the end," writes Murphy, "our government was reduced to the minimal policy of providing assistance to Hungarian refugees, and to impact on world opinion — whatever that may mean." Some 21,500 refugees were offered asylum in the US and were provided assorted governmental welfare assistance. This outcome was a long way from the pre-crisis rhetoric of "roll back," and Murphy concedes "this was not a glorious position for the United States."[35] In December 1956, Vice President Richard Nixon decried the lack of options, saying, "The United Nations has no armies that it could send to rescue the heroic freedom fighters of Hungary. There were no treaties which would invoke the armed resistance of the free nations. Our only weapon here was moral condemnation, since the alternative was action on our part which might initiate the third and ultimate world war."[36] Even more succinctly, Murphy explains, the US was "boxed."[37] One outcome of this dearth of options was the shift away from the Eisenhower Administration's emphasis on massive retaliation to the subsequent Kennedy Administration's adoption of a policy based on "flexible response." Ideally, this wider range of military options would abate the elimination by aspects limitations that hamstrung the American response to the crisis in Hungary.[38]

Example 2: Kennedy's Authorization of the Bay of Pigs, 1961

Another common heuristic is the lexicographic decision rule which assumes that each option has attributes that will promote valued outcomes. The decision-maker chooses based on the most important attribute on which a single alternative is better than all the other alternatives. When this rule is used, the chosen alternative will be viewed as dominating the rest on the attributes used in making the choice by providing the greatest utility on what has been deemed to be the most important dimension. The lexicographic strategy is agnostic of probabilities and does not require summarization of what the decision-maker knows about each option. Instead, it prescribes that one attribute be selected and the option that is best in terms of that attribute be chosen. [39]

Whereas elimination by aspects is a sequential elimination heuristic, the lexicographic rule simply involves the selection of the alternative that provides the greatest utility on what has been deemed the most important dimension.[40] In the context of the use of force, the elimination by aspects strategy would tend toward risk-avoidance as far as the desire to avoid political fallout from an ill-advised *use* of force, whereas the lexicographic model would emphasize a risk-acceptancy and the desire for political reward in terms of the rally effect *after* force is used. The elimination by aspects model is more of a defensive or "reactive" decision mechanism, while the lexicographic is offensive or "proactive."[41] This tendency of lexicographic decision-making was present in President John Kennedy's decision to authorize the Bay of Pigs operation to overthrow Fidel Castro.

Planning for the Bay of Pigs invasion began under the Eisenhower Administration. At the time Kennedy assumed office, out-going President Dwight Eisenhower had not approved the final plan but rather seemed satisfied merely "to provide his successor with a workable option for resolving the Cuba dilemma without committing American troops or adversely affecting world opinion."[42] Thus, Lucien Vandenbroucke argues, "Kennedy's first policy decision on Cuba was not to choose a course of action among the various options available. Instead, it was to decide for or against an invasion project to which considerable resources had already been committed, and that a powerful agency vigorously promoted."[43] This present study accepts Vandenbroucke's analysis and proceeds from it to depict Kennedy's decision-making as illustrative of the lexicographic rule.

Within poliheuristic decision-making, the lexicographic strategy can appear at either stage. In the first stage, it is non-compensatory, while in the second stage it is more maximizing.[44] In the case of the Bay of Pigs decision, President Kennedy emphasized the lexicographic rule during the first stage.

One of the attractions of heuristics is that they simplify the decision-making process. Kennedy certainly found himself in need of such assistance. When faced with the Bay of Pigs decision, he had been in office just seventy-seven days. As his advisor Arthur Schlesinger elaborates, Kennedy "had not had the time or opportunity to test the inherited instrumentalities of government. He could not know which of his advisors were competent and which for not."[45] In such crisis situations, Karl DeRouen and Christopher Sprecher note the particular applicability of poliheuristic theory; especially in cases like this one in which the leader sees the possibility of great political loss.[46] Mintz also notes the popularity of heuristics when time is critical. Kennedy suffered from this pressure with several advisors impressing upon him the need to make a quick decision about the Bay of Pigs because the exile invasion force could not be maintained indefinitely in Guatemala.[47]

The roots of the present crisis began in 1959, when Fidel Castro succeeded in his second attempt to overthrow the pro-American regime of Fulgencio Batista. Castro then began to implement social and economic programs designed to reduce the American influence that dominated Cuban sugar, mining, and utilities industries, including confiscating American property. He also began explaining his actions in communist language and struck a trade agreement with Russia in 1960. The United States countered by drastically reducing imports of Cuban sugar in July, but Castro responded by seizing American-owned sugar mills. A showdown in what the United States had considered its sphere of influence since the days of the Monroe Doctrine was clearly on its way.

As a presidential candidate, Kennedy was quick to make Cuba a campaign issue. Castro, he declared, was a "source of maximum danger," and his rise to power in Cuba was "the most glaring failure of American foreign policy today ... a disaster that threatens the security of the whole Western Hemisphere." Kennedy criticized Eisenhower and Vice President Richard Nixon, Kennedy's presidential rival, for allowing a "communist satellite" to appear on "our very doorstep." "I wasn't the Vice President who presided over the Communization of Cuba," Kennedy chided Nixon and reminded voters. "If you can't stand up to Castro," Kennedy asked, "how can you be expected to stand up to Krushchev?"

Kennedy criticized the Eisenhower Administration for a policy of "blunder, inaction, retreat, and failure" in Cuba and "for not doing more to remove Fidel Castro from power." "In 1952 the Republicans ran on a program of rolling back the Iron Curtain in Eastern Europe," Kennedy told voters. "Today the Iron Curtain is 90 miles off the coast of the United States." "For the present, Cuba is gone," he lamented. "Our policies of neglect and indifference have let it slip behind the Iron Curtain." If elected, Kennedy promised to reverse this trend, calling for a "serious offensive" in support of "those liberty-loving Cubans who are leading the resistance against Castro." Under Eisenhower, Kennedy claimed "these fighters for freedom have had virtually no support from our Government." Instead, Kennedy argued "the forces fighting for freedom in exile and in the mountains of Cuba should be sustained and assisted." Kennedy clearly indicated he was ready to take aggressive action. "We do not intend to be pushed around any longer,"

President Kennedy addresses the Cuban Invasion Brigade at the Orange Bowl Stadium in Miami, Florida, on December 29, 1962. From left to right in the foreground are Manuel Artime (saluting), former Cuban president José Miró Cardona, President Kennedy, and Mrs. Kennedy (photograph by Cecil Stoughton; John F. Kennedy Presidential Library and Museum).

he crowed, "and in particular do not intend to be pushed out of our naval base at Guantanamo, or denied fair compensation for American property that has been seized."[48]

Kennedy's campaign rhetoric notwithstanding, Eisenhower certainly had not been ignoring developments in Cuba. Discussions of a program designed to oppose Castro began as early as October 1959, and in January 1960, the CIA established a Task Force WH-4 (Branch 4 of the Western Hemisphere Division) to implement Eisenhower's request for an ambitious covert program to overthrow the Castro government. Reflecting the influence of the CIA's successful ouster of Guatemalan President Jacobo Árbenz Guzmán in 1954, Jacob Esterline, the Guatemala station chief between 1954–1957, was put in charge of WH-4.[49]

Official sanction for the plan came on March 17 when Eisenhower reports he "ordered the [Central Intelligence] Agency to commence the training of Cuban exiles, principally in Guatemala, for the possible future day when they would return to their country."[50] A CIA policy paper titled "A Program of Covert Action Against the Castro Regime" outlined four main themes: creation of a unified political force of opposition to Castro outside Cuba, development of a propaganda program, creation of an underground force on Cuba and the promotion of insurgency in the country's mountains, and creation of a paramilitary force outside Cuba.[51]

The lexicographic rule holds that a decision-maker bases his decision on an alternative's adherence to a single most important attribute. Kennedy's bellicose stand on Cuba during the campaign may have helped him win a tight presidential election, but it also "painted him into a political corner" and made political considerations the decisive attribute in his decision-making process about the Bay of Pigs invasion.[52] In his restrained and patient approach to Cuba, Eisenhower had the benefit of the prestige and credibility of have been the Supreme Allied Commander during World War II. In contrast, Kennedy was "a young and unproven leader with little foreign-policy experience."[53] In addition to his bellicose campaign rhetoric, Kennedy had promised in his inaugural address to "pay any price, bear any burden, meet any hardship" to defend freedom. Of specific relevance to the situation in Cuba he said, "Let all our neighbors know that we shall join with them to oppose aggression or subversion anywhere in the Americas, and let every power know that this hemisphere intends to remain the master of its own house."[54]

It was Kennedy's strong anti–Castro rhetoric that leads Vandenbroucke to posit, "perhaps the single most important reason behind [Kennedy's Bay of Pigs] decision was the desire to avoid domestic criticism on the issue of communism."[55] This consideration certainly appears to have transcended reservations Kennedy had about the operation. "I'm still trying to make some sense out of it," he confided to Schlesinger just days before the invasion.[56] Others counseled caution. Senator William Fulbright, influential chairman of the Senate Foreign Relations Committee, argued, "The Castro regime is a thorn in the flesh, but not a dagger in the heart," and even the generally-hawkish Dean Acheson called the plan "a wild idea."[57]

Numerous observers, however, believe political considerations trumped these concerns. Michael Grow asserts, "Domestically, political considerations virtually forced Kennedy to pursue an interventionist policy."[58] Robert Divine concludes "Kennedy went ahead, aware that he had committed himself to a strong course of action by his criticism of Ike's failure to stop Castro."[59] Deputy National Security Advisor Walter Rostow agrees. "Kennedy had taken an activist position during the campaign," Rostow explains. If, as president, Kennedy had failed to execute Eisenhower's plan, "the Republicans could have argued that Kennedy didn't have the guts to go through with something that would have eliminated Castro once and for all. Everything unpleasant that happened subsequently in Cuba would have been directly on Kennedy."[60] Likewise, Vandenbroucke writes, "despite serious misgivings about the plan, [Kennedy] felt nevertheless compelled to accept it, largely out of fear of domestic criticism."[61]

Such a response is consistent with the lexicographic rule's expectation that the decision-maker will select the alternative that provides the greatest utility on what he considers the most important dimension. The result is a risk-acceptant or offensive approach in an effort to secure political rewards. In Kennedy's case, the value he attached to consistency with his tough campaign rhetoric outweighed all other considerations, to include the probability of the operation's success.

The order in which information is presented also affects decision-making.[62] Kennedy's decision-making process did not begin from a neutral position in which various courses of action could then be developed based on his guidance. Instead, he was presented with a plan that was already in existence and asked to decide on it. Vandenbroucke writes Kennedy "suddenly faced a plan that matched his campaign rhetoric."[63] He likely felt such a de facto situation narrowed his options.

Having decided in the first stage to go ahead with an invasion, Kennedy then had to decide on several operational details. As he weighed various alternatives, his primary concerns during the second stage of his decision-making process were to avoid charges of US involvement and to reduce the visibility of the operation.[64] Based on these criteria, Kennedy changed the landing site from Trinidad to the Bay of Pigs and canceled plans for a second air strike.[65] While such decisions reflect some characteristics of the rational actor model, their sequencing clearly places them within the poliheuristic procedure. As Vandenbroucke notes, "the president and his advisors did not carefully weigh competing alternatives and then select the invasion of Cuba as the best policy."[66] Instead, Kennedy used a lexicographic strategy in the first stage of his decision-making process to commit himself to an invasion. Then, in the second stage, he considered various alternative forms of that invasion.

Political considerations remained critical during the second stage. The decision to change the landing site was made in spite of Trinidad's many military advantages, and the fact that "other possible landing points had been examined, but each had been lacking in some vital component." Trinidad had a harbor, a defensible beachhead, a remoteness from Castro's main army, and easy access to the Escambray Mountains as a safe haven in the event of a defeated operation. Central Intelligence Agency agent Grayson Lynch, who participated in the invasion, declared, "Trinidad had everything the planners were looking for." In his mind, the site was "ideal." The Joint Chiefs of Staff also favored Trinidad, but Kennedy considered the plan there "too spectacular." He emphasized he did not want anything reminiscent of the large-scale World War II-type amphibious operations that might attract undesirable international attention, and instead wanted a "quiet" landing, preferably at night.[67]

Based on Kennedy's Trinidad apprehensions, the CIA came up with three new landing sites: the Preston area on the north coast of Oriente Province, the south coast of Las Villas between Trinidad and Cienfuegos, and the eastern Zapata area near Cochinos Bay (or the "Bay of Pigs").[68] Based on political considerations, Kennedy selected the Bay of Pigs, some hundred miles west of Trinidad. From a military operational perspective, the Joint Chiefs of Staff thought the Bay of Pigs was the best of the three new alternatives, but that none of the three was as likely to accomplish the mission as Trinidad.[69]

Kennedy also made the decision to cancel the planned second air strike in the second stage of the poliheuristic process. After repeatedly asking CIA Deputy Director of Plans Richard Bissell, "Do you really have to have these air strikes?" Kennedy decided on April 14 that he wanted only a "minimal" level of activity and ordered the number of planes involved in the D-1 airstrikes scheduled for April 15 to be reduced from sixteen to six.[70] Then at about 9:30 P.M. on April 16, National Security Adviser McGeorge Bundy telephoned Major General Charles Cabell, Deputy Director of the CIA, informing him the dawn airstrikes scheduled for the following morning should not be launched until it was reasonable to assume planes could conduct them from a strip within the beachhead. Prior to that, public knowledge of the flights' origination from Nicaragua would raise unacceptable international attention, and Kennedy decided he was "not signed on to this."[71] When Cabell and Bissell protested to Secretary of State Dean Rusk, Rusk said he had just recommended to Kennedy and Kennedy had agreed to cancel the D-Day strikes. [72] Bissell considered the D-Day strikes critical to the safety of the landings and wrote in his memoirs that he believed "the president did not realize that the air strike was an integral part of the operational plan he had approved."[73] Indeed, Kennedy was developing his own doubts. He later confided to advisor Theodore Sorenson that he wished he had canceled the entire operation rather than just the second airstrike. Sorenson reports, "It was clear to [Kennedy] by then that he had in fact approved a plan bearing little resemblance to what he thought he had approved."[74]

Sorenson attributes much of the Bay of Pigs' failure to the process having been governed by "bureaucratic momentum" rather than "policy leadership."[75] Some of the "momentum" to which Sorenson alludes can be credited to the two-stage poliheuristic procedure. Having committed himself to an invasion in the first stage, Kennedy then continued to make suggestions "to reduce the noise level" as the process of finalizing the plan proceeded through the second stage.[76] During this process, Schlesinger interprets that Kennedy understood the dynamic as being "the smaller the political risk, the greater the military risk, and vice versa." "The problem was to see whether the two risks could be brought into balance."[77] Unfortunately, Schlesinger concludes, "given the nature of the operation, this was impossible, and someone should have said so."[78] From the perspective of poliheuristic theory, however, once Kennedy made the decision to conduct an invasion in the first stage of the process, the time for discussions of the sort Schlesinger suggests had passed. By then, Schlesinger declares Kennedy and his advisors to have become "prisoners of events."[79] At some point during the second stage, Secretary of State Rusk mused, "Maybe we've been oversold on the fact that we can't say no to this," but in the sequential poliheuristic approach, first stage decisions are seldom revisited.[80]

As is often the case with such a methodology, the results were disappointing. Indeed the April 17 invasion proved to be Kennedy's "worst hour."[81] The anticipated spontaneous uprising of the Cuban people failed to materialize, and without air support, the 1,500-man exile force was pinned down on the beaches. Within two days Castro had eliminated the threat, taken over a thousand prisoners, and humiliated Kennedy and the US.[82] In retrospect Kennedy wondered, "How could I have been so stupid to let them go ahead?"[83] At least part of the answer lies in the dangers of the less than optimal choices often generated by poliheuristic decision-making.

Example 3: Johnson's Decision
to De-escalate US Involvement in Vietnam, 1968

Decision heuristics such as elimination by aspects and the lexicographic rule are part of the cognitive psychology represented in the first stage of poliheuristic decision-making. The second stage uses an analytic process that corresponds to rational choice theory. To help understand the entire poliheuristic process, Mintz outlines a procedure to help dissect how a leader makes a decision under such circumstances. Mintz's model can be used to analyze President Lyndon Johnson's decision-making process in the aftermath of the Tet Offensive during the Vietnam War.

By late 1967, many American generals believed that their relentless search and destroy missions were wearing down the enemy by the steady process of attrition. The communists seemed to share this assessment that the momentum lay with the Americans. Although they had been prepared to take large numbers of casualties to achieve their victory, the North Vietnamese leadership feared that their recent losses had pushed them to the breaking point. Therefore, in January 1968, they deviated from their strategy of protracted war and launched the massive Tet Offensive aimed at creating a popular uprising among the South Vietnamese people. Instead, the offensive turned into a decisive defeat for the communists, leaving the US with a decision to make regarding its next move.

The communist plan called for a three-stage offensive. The first phase was to lure the American forces away from the cities of South Vietnam by drawing them into battles in the remote corners of the country. Of particular importance was an North Vietnamese Army buildup around the Marine base of Khe Sanh. With American forces focused elsewhere, the Viet Cong would initiate the second stage by secretly moving an assault force of about 85,000 men into position for a country-wide attack. The third stage was the massive simultaneous assault on the major cities of South Vietnam.

In spite of the initial advantage of surprise, the Tet Offensive proved to be a military disaster for the communists. They suffered huge losses, with battle deaths estimated as high as 40,000. Perhaps even more significantly, the South Vietnamese people had failed to rise up against their government as the communists had expected. Instead, the Viet Cong infrastructure that had taken years to build was destroyed almost overnight.[84]

For the United States and South Vietnam, Tet represented their greatest victories of the war. In previous fighting, the US had been unable to apply its superior firepower against an enemy that refused to engage in decisive battle. Deviating from guerrilla tactics and fighting a conventional battle, the communists exposed themselves to the full weight of the US military. Moreover, the South Vietnamese Army (ARVN) showed a previously unrealized resolve and acquitted itself well in this moment of crisis.[85] Military Advisory Command, Vietnam commander General William Westmoreland felt confident that he could follow up on this victory with a more aggressive strategy that "could materially shorten the war."[86] To take advantage of this military opportunity, Westmoreland and Chairman of the Joint Chiefs of Staff General Earle Wheeler asked President Johnson for an additional 206,000 troops. Many of the troops would be earmarked to shore up American strategic commitments worldwide, but in Vietnam, the extra manpower would allow a more aggressive strategy that included expanding the war into Laos, Cambodia, and possibly that part of North Vietnam just above the demilitarized zone.[87] In considering this request, Johnson acted in a manner consistent with the predictions of the poliheuristic theory.

Applied decision analysis (ADA) is an analytic procedure designed to recreate or "reverse engineer" a particular decision-making process. It is an effort to "enter the minds of decision makers in an attempt to uncover their decision rules."[88] Mintz lays out a two-step ADA process that can be used to apply the poliheuristic procedure to help understand how a leader like Johnson makes a decision. The first is to identify the decision matrix used by the leader. This requirement includes identifying the policy alternatives available, identifying the dimensions or criteria that are relevant in evaluating the alter-

General William Westmoreland briefs President Lyndon B. Johnson and his advisors in a Cabinet Room meeting on April 27, 1967. From left to right are Walt Rostow, General Westmoreland, Nicholas Katzenbach, Secretary of State Dean Rusk, President Johnson, Cyrus Vance, and General Earle Wheeler (LBJ Library, photograph by Frank Wolfe).

natives, understanding the implications each alternative has on each dimension, determining how the implications are rated, and determining how each dimension is weighted to indicate its level of importance. The second step is to apply poliheuristic calculations to the decision matrix. This step involves determining which alternatives with very negative values on the political dimension to discard and then evaluating the remaining alternatives based on rational calculations.[89]

President Johnson's decision matrix had at least four policy alternatives that were presented to him by his staff on February 27: comply with Westmoreland's request, combine the military increase with a new peace initiative, maintain the status quo on troop commitments and develop a new strategy, or commit "a huge number of men — 500,000 to a million more." The last alternative was presented by incoming Secretary of Defense Clark Clifford as merely a possibility. He did not advocate it, and no one else at the meeting appeared to champion it either.[90] In a practical sense then, Johnson had to choose between two alternatives that increased the US commitment and one that did not. In use of force scenarios such as this one, the relevant dimensions typically are military, economic, political, and diplomatic consequences.[91] Within each of these dimensions, Johnson had to determine the implications for each alternative.

The implications associated with the military dimension created considerable consternation among the members of the Johnson Administration. In order to comply with the request, Johnson would have to mobilize the National Guard and the Reserves, something he had not done to this point. Several cabinet members opposed this move, including Secretary of State Dean Rusk and departing Secretary of Defense Robert McNamara. Economically, McNamara estimated the new troop commitment and call up of reserves would increase the budget by $2.5 billion in 1968, by $10 billion in 1969, and by $15 billion in 1970. Such numbers were sure to cause inflation, a large tax increase, or both in the midst of an already strained budget.[92] Indeed, when word of the Westmoreland's troop increase proposal leaked out, there was a rush to trade dollars for gold in financial markets throughout the world.[93] Among the more serious diplomatic implications was the damage to America's credibility in the eyes of the world if it failed in Vietnam.[94] Soviet and Chinese responses to any US action also had to be considered.

The greatest implications lay within the political dimension. Westmoreland had returned to the US in November 1967 and gone on a highly orchestrated, whirlwind speaking tour to reassure the American public, telling them that "we have reached an important point when the end begins to come into view." "We are making real progress," Westmoreland assessed, and he predicted "turning the war over to the South Vietnamese in less than two years."[95] Such optimistic predictions came back to haunt Westmoreland after the communists launched Tet. On February 6, 1968, Art Buchwald lampooned in the Washington *Post*: "Little Big Horn, Dakota, June 27, 1876. General George Armstrong Custer said today in an exclusive interview with this correspondent that the Battle of Little Big Horn had just turned the corner and he could now see the light at the end of the tunnel. 'We have the Sioux on the run,' Gen. Custer told me. 'Of course, we still have some cleaning up to do, but the Redskins are hurting badly and it will only be a matter of time before they give in.'"[96]

The extensive television news coverage of the Tet Offensive did not seem to depict the inevitable victory Westmoreland had described. If the enemy was near defeat, how could it launch such a massive, country-wide attack and how could it catch US and South Vietnamese forces so unaware? Key to the shaping of public opinion was *CBS Evening News* anchor Walter Cronkite who traveled to Vietnam to cover the Tet story. On February 27, he reported, "The bloody experience of Vietnam is to end in stalemate ... it is increasingly clear to this reporter that the only rational way out [of the war] will be to negotiate, not as victors, but as an honorable people who lived up to their pledge to defend democracy, and did the best they could." In the age before cable news, network newsmen were enormously influential, and Cronkite personally was credited with being "the nation's most trusted person." After Cronkite's reversal on the war, Johnson lamented, "If I've lost Cronkite, I've lost Middle America."[97]

As Johnson rated these implications, all would be in the negative domain. The differences were only in degrees. On the economic front, Johnson confides that "monetary and budgetary problems were constantly before us as we considered whether we should or could do more in Vietnam." At a March 15 staff meeting, Rusk cautioned that a dramatic military increase could set off a panic in Europe, heavy domestic inflation, and even a collapse of the monetary system. "If we do this without a tax bill," Rusk warned, "we are dead."[98] Moreover, new military costs revived the "guns or butter" debate and posed a serious threat to Johnson's "Great Society" programs. In this context, mounting war costs "were given special pertinence by the rioting in the urban black ghettoes in 1967 and 1968 and by the deterioration of American cities which these racial explosions held up for all to see."[99] The costs of Vietnam and the Great Society combined to produce an estimated 1968 budget deficit between $25 and $30 billion.[100] Indeed, Johnson highlights "our financial problems" as playing "an important part" in his post–Tet decision-making.[101]

Militarily, even out-going Secretary of Defense McNamara refused to endorse what would amount to a rise in the total military troop strength to 731,756 men.[102] To meet the Westmoreland request, McNamara anticipating having to call-up about 250,000 reserves from all services and having to extend enlistments by six months. Johnson was especially sensitive to the politically charged issue of reserve call-ups, remembering complaints generated by call-ups made by President John Kennedy and the more recent failure to use effectively reservists called up during the USS *Pueblo* crisis. Before considering any reserve call-ups, Johnson wanted to know:

> Why ... is it necessary to call up reserve units at this time? If we decided on a call-up, how large should it be? Could we reduce the numbers by drawing on forces stationed in Europe or South Korea? Could we avoid or at least postpone individual reserve call-ups? If reserves were called, where would they be assigned? How long would they serve? What would be the budgetary implications? Would congressional action be necessary?[103]

Secretary of State Rusk captured the diplomatic implications, telling Johnson, "The integrity of the US commitment is the principal pillar of peace throughout the world. If that commitment becomes unreliable, the communist world would draw conclusions that would lead to our ruin and almost certainly to a catastrophic war."[104] Similarly, Rusk's undersecretary George Ball noted, "the authority of the United States in the world depends, in considerable part, on the confidence of other nations that we can accomplish whatever we undertake."[105] In a 1965 speech at Johns Hopkins University, Johnson had said "to leave Vietnam to its fate would shake the confidence of all [the world's] people in the value of an American commitment and in the value of America's word."[106]

The present US hold on this confidence already was tenuous, and Johnson assessed there being "many sources of anxiety and many unknown factors" involving American commitments to its allies. The situation in Korea was "explosive," with President Chung Hee Park having only recently survived an assassination attempt by North Korean operatives and the North Korean seizure of the *Pueblo* still unresolved.[107] To make matters worse, intelligence reports indicated "a crisis might develop around West Berlin."[108] Johnson was counseled by advisors such as General Matthew Ridgway that the US strategic reserve was stretched dangerously thin to respond to any new situation that might erupt outside of Vietnam.[109] When advisors debated whether a declaration of war would be necessary to implement the reserve call-ups needed to preserve the strategic reserve capability, Rusk warned a declaration would represent "a direct challenge to Moscow and Peking in a way we have never challenged them before" and might have a "very severe" international impact.[110]

The political implications were the worst. Rather than the "rally round the flag" effect presidents often enjoy in many such situations, Johnson suffered a sharp drop in public approval ratings, and he recognized that there had been "a dramatic shift in public opinion on the war."[111] Indeed, Gallup poll data suggest that between early February and mid–March 1968, nearly one person in five changed from the "hawk" to the "dove" position.[112] Johnson's job ratings plummeted to new lows with twenty-six percent of those polled approving of his handling of the war and sixty-three percent disapproving. Thirty-six percent gave Johnson overall approval, but fifty-two percent were negative.[113]

In 1966, Johnson had asserted, "I'm not going down in history as the first American President who lost a war." After Tet, he found himself in an even more precarious position. Following Mintz's option of assigning numerical ratings to the implications, if -10 is very bad and + 10 is very good, Johnson might have been facing something like a -8 politically and somewhere between a -5 and a -3 in each of the other three areas.

The final step according to the poliheuristic procedure of understanding the decision matrix of the leader is to assign "weights" to the importance level of each dimension. Mintz suggests using a scale from 0 (not important at all) to 10 (very important). Clearly, Johnson would weigh the political dimension the most heavily, placing it far on the "very important" end of the scale.

Poliheuristic theory posits that after conducting this analysis, decision-makers will next eliminate from consideration alternatives that are harmful to the leader. Specifically, the non-compensatory principle will lead decision-makers to avoid major political loss.[114] Thus alternatives that have a very negative value on the political dimension will be discarded first. Applying this logic, Johnson could readily see that the American public would not support Westmoreland's requested troop increase. To this end, historian Lieutenant General Dave Palmer goes as far as to say "the public rebelled" against widening the war. "In the election year of 1968 there could be no further escalation of the conflict," Palmer continues, "not if the Democrats hoped to retain the White House."[115] As poliheuristic theory would predict, Westmoreland's proposal, with its prohibitively negative political implications, would be discarded.

This decision was consistent with the non-compensatory notion that "a low score on one dimension cannot be compensated by for by a high score on another dimension."[116] From a military point of view, "as the dust settled and Allied generals began to realize the full consequences of the Tet Offensive, they became in many ways more optimistic than ever."[117] Westmoreland thought it was time to "consider a change in strategy, take advantage of the enemy's having come into the open, and strike boldly, breaking away from the long war of attrition that political restraints had for so long imposed."[118] As the poliheuristic theory would predict, however, for Johnson no potential opportunity in the military dimension could offset the negative impact on the political dimension. "We need more money in an election year, more taxes in an election year, more troops in an election year and more [budget] cuts in an election year," Johnson moaned. The result was "we have no support for the war."[119]

Before making a decision on the troop request, Johnson charged Clifford, who Johnson felt had "a new pair of eyes and a fresh outlook," to study the situation. Five days later, the new secretary of defense recommended against granting Westmoreland's request. Johnson readily accepted Clifford's recommendation, based on reasoning that appears consistent with the poliheuristic theory. Johnson's biographer Robert Dalleck explains, "An increase in US force levels proposed by the Chiefs seemed likely to produce a fiscal-economic crisis and a renewed assault on Johnson's credibility, which would further weaken his political hold on the country."[120] Thus instead of following the course of Westmoreland and Wheeler to seize the military opportunity presented by Tet to deal the communists a lethal blow, the machinations of the poliheuristic theory turned the victory on the battlefield into a crushing moral defeat.[121] Having eliminated this politically infeasible solution, Johnson's next task was to select from the remaining alternatives "the alternative that has the best net gain on all dimensions (or on the dimension most important to the decision-maker)."[122]

Continuing with the present course seemed to offer little promise. As Clifford recalled from his investigation into Westmoreland's request for additional troops,

> I could not find out when the war was going to end: I could not find out the manner in which it was going to end. I could not find out whether the new requests for men and equipment were going to be enough, or whether it would more and, if more, how much. All I had was the statement, given with too little self-assurance to be comforting, that if we persisted for an indeterminate length of time, the enemy would choose not to go on.[123]

As an alternative, Clifford recommended and Johnson quickly accepted a proposal to improve and modernize the equipment of the ARVN and develop a program to gradually shift the burden of the war effort to the South Vietnamese.[124] Under President Richard Nixon, this strategy would become full-scale "Vietnamization" and ultimately lead to a US withdrawal from the war. For the duration of Johnson's term, it became what George Herring describes as a strategy of "fighting while negotiating." As such, it was an admission of strategic failure.[125]

Poliheuristic decision-making presents the danger of producing suboptimal solutions that settle for something short of the singular national security focus of decisions made pursuant to the rational actor theory. Indeed, all Johnson told Clifford that he wanted was "the lesser of evils."[126] Setting aside any discussion of the merits of the US involvement in Vietnam and the strength of its connection to US national security, from the decision-maker's point of view — in this case that of President Johnson — Vietnam was a war America needed to win or at least end with a peace that preserved South Vietnam.[127] Westmoreland writes in his memoirs that the war "could have been brought to a favorable end following defeat of the enemy's Tet offensive ... had President Johnson changed our strategy and taken advantage of the enemy's weakness."[128] Not all observers agree with Westmoreland's assessment, but for the purposes of this discussion, the point is that the call-up of reserves and the new strategy was rejected not for military reasons, but because it was "politically infeasible."[129] Johnson concluded that he would have to forego winning the war if doing so required a politically unacceptable military escalation.[130] Consistent with the non-compensatory principle, Johnson eliminated the escalation option, and instead settled for a suboptimal solution that was tantamount to admitting the war was unwinnable.[131]

Herring clearly identifies the poliheuristic component and shortcoming of Johnson's course. Rather than solving the root problem, Johnson's decision "appears to have been designed to quiet the home front as much as anything else." As a result, Johnson was left "picking and choosing from conflicting approaches rather than forming a coherent strategy to achieve a clearly defined objective."[132] The contradictions inherent in the decision were problematic. As Herring explains,

> Militarily, the means were scaled back without modifying the ends, and it is impossible to see in retrospect how US officials hoped to achieve with the application of less force aims that had thus far eluded them. The United States had taken over the war in the first place because of the poor performance of the South Vietnamese, and the concept of Vietnamization was equally illusory. Negotiations were desirable from a domestic political standpoint, but in the absence of a military advantage the United States did not have and concessions it was not prepared to make, they could not achieve anything. And the mere fact of negotiations could soften US resolve and limit the administration's ability to prosecute the war.[133]

Instead of resolving the war one way or the other, all Johnson's post–Tet decision did was perpetuate the stalemate — which had been the problem his decision theoretically had set out to address.[134]

Utility of the Poliheuristic Theory

Because it uses decision shortcuts and rules of thumb, poliheuristic theory can explain complicated foreign policy decisions, and it is compatible with a host of contingency theories that attribute to the decision-maker requisite flexibility to adapt the decision process to changing problems and conditions.[135] A major step in extending poliheuristic theory would be to automate the theory and its decision rules, thereby making it easier for scholars to apply and test the theory using different data sets in a variety of geographical and historical contexts. As a step in this direction, Mintz has launched a "computerized decision process tracing" process called "Decision Board." It can be accessed at http://www.decisionboard.org/academic/zzzindex.asp. This software can be used "to develop and organize the information relevant to individuals' decisions through the construction of decision boards and matrices."[136]

Another advancement would be to extend the poliheuristc model to group decision making and sequential decision making. Decision processes and choices are shaped by the "decision unit" or who makes the decision.[137] The simplest representation of poliheuristic theory is as a single individual making choices based on the non-compensatory decision principle. However, political choices in bureaucratic or democratic settings are often the product of group and societal processes which require an individual leader to interact with others to make and implement choices.[138] The decision-making process of a single individual should not be expected to follow the same pattern as a decision made by a group.[139] However, poliheuristic theory has been criticized for appearing to be limited to decisions in which a single predominant leader has the ultimate authority to choose a course of action. Such a construct relegates organizational members to an exclusively advisory role when in fact decisions may involve the input of a number of individuals whose preferences must be satisfied, depending on the institutional structure of the decision making unit.[140] Extending poliheuristic theory to accommodate these situations would expand its applicability, but in the meantime, the bureaucratic model may be a more useful analytical tool.

4

The Bureaucratic Model

The bureaucratic model (also called the bureaucratic politics, government bargaining, governmental, or institutional pluralism model) argues that decisions result from the bargaining process among various government agencies that have somewhat different interests in the outcome.[1] This model rests on four key assumptions. The first is that "policy-making is a social process." Although a certain leader like a president may have ultimate responsibility for the policy, a multiplicity of individuals and institutional actors play a role in the decision-making process. The second assumption is that "there is no single version of the national interest." While there may be some general consensus on higher order issues, each participant also has his own set of perspectives and interests shaped by his bureaucratic role. In his famous analysis of the Cuban Missile Crisis, Graham Allison captured this reality, explaining, "Where you stand depends on where you sit."[2] The third assumption is "policy decisions are compromises." This assumption follows from the first two as the various actors attempt to resolve their different perspectives and frames of reference. Thus decisions will not necessarily reflect a cooperative effort to achieve a unitary goal but rather are the compromise that emerges as the result of competition. The final assumption is "politics does not stop once a decision is made." Because the participants surrendered some of their preferences in the compromise, they will often try to regain ground during the implementation of the decision by tactics such as foot-dragging, adhering to the letter but not the spirit, or lack of follow-through.[3]

The result is that the bureaucratic model views decision-making as a decentralized process conducted in an abstract political space rather than a formal procedure that relies on a hierarchical chain of command. Rather than seeking a unified effort directed toward a common objective, the bureaucratic organizational heads jealously guard their agency's area of expertise. In the process, they may use strategies such as withholding information or negotiating internally to limit options available to the decision-maker to those they believe represent their organization's best interest. As a result, the bureaucratic model often results in fragmented decision-making.[4] The bureaucratic model is present in the departmental system used to organize the defense of the Confederacy, the debate between the Army and Air Force over helicopter proponency, and the conflicting views of Secretary of State George Shultz and the Secretary of Defense Caspar Weinberger regarding the insertion of a peacekeeping force in Beirut.

Example 1: The Confederate Departmental System's Impact on the Vicksburg Campaign, 1862–1863

The Confederacy had a vast amount of territory that needed to be somehow organized militarily, but the South's strong adherence to the principle of state's rights impeded efforts to form an efficient, centralized command system. The result was a departmental organization of regional commands which were largely founded on state lines and geographic features. Such an organization lent itself to efficient peacetime administration, but it also led to excessive reliance on the bureaucratic model which often failed to produce decisions that provided the unity of effort and flexibility needed in wartime.[5] One specific example of the departmental system leading to the use of the bureaucratic model is the Vicksburg Campaign.

Under the departmental system, each department was commanded by an officer of appropriate rank, and most operational decisions were left to these department commanders. Theoretically, this arrangement would allow the Confederate government to focus on only the most important strategic decisions. In reality, the department commanders tended to operate in isolation from each other with only little inter-department cooperation.[6] They naturally viewed their own problems as the most critical ones facing the country.[7]

The result of this organizational technique was that policy decisions reinforced the bureaucratic model's assumption that "there is no single version of the national interest." Because each department commander reported directly to President Jefferson Davis, their sense of autonomy tended to make them think almost exclusively in terms of their own regional responsibilities. Too often, the preservation of autonomous departmental organization came at the expense of inter-departmental cooperation. As Captain R. G. H. Kean, chief of the Confederate Bureau of War, saw the situation, "the inviolability of a departmental line" caused the "separate departmental organization" to become the "radical vice of Mr. Davis' whole military system."[8]

By delegating authority and resources to the department commanders from the outset, President Davis left himself very little ability to later influence the situation. He was also reluctant to go against the judgment of a local commander in whom he had entrusted so much authority. Therefore, the system was based on a tremendous reliance on the department commanders—some of whom warranted such trust while others did not. For this reason, much of the military history of the Confederacy is biographical, being highly influenced by the personalities of men like Generals Robert E. Lee, Braxton Bragg, and Joseph Johnston. According to Thomas Connelly and Archer Jones, "President Davis probably never realized the degree to which his commanders developed ... local attachments and perspectives."[9] As a result, Davis, like other presidents who rely on this decision-making model, found himself "reined in by the bureaucracy" rather than served by it.[10]

The prevalence of the departmental system had strategic implications, in many ways becoming an end unto itself, with President Davis shaping "policy to fit the needs of the departmental system, rather than the reverse."[11] In fact, because Davis and his War Department were reluctant to order cooperation, the departmental system served to preclude any effective means of strategic direction.[12] Eugene Wittkpof and his colleagues identify this ad hoc decision-making as one of the characteristics of the bureaucratic model.

"Rather than choices being made in light of carefully considered national goals and long-term planning," Wittkopf explains, "trial-and-error responses to policy problems as they surface seem more characteristic."[13] Indeed, Kean reported President Davis seemed "to have no plan — to be drifting along on the current of events."[14] While Thomas Connelly and Archer Jones more sympathetically conclude Davis improved in his use of the departmental system as a strategic tool, even they admit he merely "moved from crisis to crisis."[15] Such an outcome is consistent with arguments that the key characteristic of the bureaucratic model is "that there is no overarching master plan."[16] Instead it tends to react to demands rather than to priorities.

One of the main manifestations of this lack of a comprehensive Confederate strategy was the difficulty in achieving the Jominian principle of concentration. At the beginning of 1863, the Confederacy was faced with two completely different situations in the eastern and western theaters. In the east, Lee's spectacular victory at Chancellorsville had left the Army of the Potomac reeling and presented the opportunity for the Confederates to launch a major offensive. In the west, the situation was reversed with Major General Ulysses Grant posing a mounting threat to Lieutenant General John Pemberton's beleaguered command at Vicksburg and the Confederacy's ability to maintain some control over the strategically important Mississippi River.[17] On May 19, Secretary of War James Seddon, encouraged by the likes of General Pierre Goustave Toutant Beauregard, Lieutenant General James Longstreet, and Senator Louis Wigfall, asked Lee for his thoughts on sending one of his divisions west.[18] For Lee, the discussion boiled down to "a question between Virginia and the Mississippi."[19] Based on his position within the Confederate bureaucracy, Lee had little trouble advising President Davis which to choose.

Simply put, Connelly and Jones summarize that "Lee's strategy was based on defending Virginia."[20] In advancing this cause, Lee benefited from his exalted standing within the Confederacy and with President Davis, who quickly concurred with Lee's analysis. Alex Mintz and Karl DeRouen note that within the bureaucratic model, "personalities and agency clout become deciding factors," and "larger-than life figures ... have inordinate power."[21] Certainly, Lee's prestige gave him an advantage in a model in which Graham Allison argues "each player's ability to play successfully depends on his power." Allison goes on to describe power as "an elusive blend of at least three elements: bargaining advantage (drawn from formal authority and obligations, institutional backing, constituents, expertise, and status), skill and will in using bargaining advantages, and other players' perceptions of the first two ingredients."[22] In all these areas, Lee excelled his competitors, giving him valuable leverage in what Allison calls the "pulling and hauling" of bureaucratic politics.[23] According to Jones, "the prestige of General Lee, thrown unequivocally and forcefully behind a proposal to continue to rely solely on the departmental system, must have been overwhelming" in its influence on President Davis.[24]

As the bureaucratic model would predict, the departmental system's emphasis on self-interest caused some commanders to feel that if they lent forces to another department in response to a threat there, they were dangerously exposing their own department to attack. Instead of shifting forces from one department to another, some argued that they could best assist a threatened department by going on the offensive in their own department. This concept was based on the idea that the new offensive would compel the Federals to release forces from in front of the originally threatened department in order to meet the new development. This was the logic Lee used to support his Gettysburg

Campaign instead of sending reinforcements to Vicksburg, arguing "greater relief would in this way be afforded to the armies" threatened by a Federal advance "than by any other method."[25] While Lee may have in fact believed such would be the case, more detached observers attribute the primary motivation of his recommendation to have been his self-interest in the eastern theater. Indeed, one colleague concluded "Robert E. Lee regarded his allegiance to the sovereign state of Virginia as paramount to all others."[26]

Lee had elsewhere advocated an "offensive-defensive strategy," explaining, "It is [as] impossible for [the enemy] to have a large operating army at every assailable point in our territory as it is for us to keep one to defend it. We must move our troops from point to point as required, and by close observation and accurate information the true point of attack can generally be ascertained."[27] However, his "concern for Virginia, his conviction that it was to be the principal point of attack, and his lack of knowledge of the West led him to advocate ideas on grand strategy which were basically foreign to his own perceptions of strategy."[28] Thus when Lee argued against reinforcing Pemberton at Vicksburg, Archer Jones believes the Virginian "seemed to suffer from cognitive dissonance [the psychological discomfort resulting from attempting to reconcile knowledge that contains contradictions] in trying to reconcile his allegiance to his army with his ingrained strategic concepts."[29] Lee was not alone. Connelly and Jones conclude "frequently a general preached concentration and mutual support until he gained his own department, then became overpossessive and too cautious, even selfish." They cite Generals Pierre Gustave Toutant Beauregard and Joseph Johnston as examples.[30]

The decision having been made to focus on the eastern theater, the departmental system further hamstrung Pemberton in the west. The western boundary of his department rested upon the Mississippi River, the largest high-speed avenue of approach in North America, and Pemberton had no authority over forces on the far shore. Such a condition led to bureaucratic decision-making that fractured unity and "ensnared the Confederacy's river defense in command squabbles."[31]

As early as November 1862, General Samuel Cooper, the Confederate Army's Adjutant General, had been bombarding Department of the Trans-Mississippi commander Lieutenant General Theophilus Holmes with requests such as "Can you send troops from your command — say 10,000 — to operate either opposite to Vicksburg or to cross the river?" Holmes parried each request, complaining that to comply would threaten Arkansas. Eventually Cooper acquiesced, telling Holmes, "you must exercise your judgment in the matter." President Davis resumed the dialogue later in December, writing Holmes that it was "unquestionably best" that Holmes reinforce his neighbor department east of the Mississippi River.[32] However, Davis stopped short of ordering Holmes to do so; instead functioning "through a laissez faire management style of his departmental system."[33] As a result, critics accuse Davis of "passing the buck to departmental commanders" rather than taking responsibility as chief executive.[34] Such "decision avoidance" is another danger Wittkopf identifies as being characteristic of the bureaucratic model.[35]

Lieutenant General Edmund Kirby Smith superseded Holmes in February 1863, and on May 9, Pemberton advised his new counterpart, "You can contribute materially to the defense of Vicksburg and the navigation of the Mississippi River by a movement upon the line of communications of the enemy on the western side of the river....To break this would render a most important service." Finally, Major General John Walker's Texas Division began operating on the east side of the Mississippi and attacked Milliken's Bend on June 7. Even then, support from the Trans-Mississippi was grudging. Smith's independent-minded subordinate, Major General Richard Taylor, who commanded the District of Western Louisiana, preferred to use Walker against New Orleans, but was overruled by Smith. Still Taylor complained in his memoirs that "remonstrances were to no avail. I was informed that all the Confederate authorities in the east were urgent for some effort on our part in behalf of Vicksburg, and that public opinion would condemn us if we did not *try to do something*." He insisted "that to go two hundred miles and more away from the proper theatre of action in search of an indefinite something is hard; but orders are orders." The critical Kean was aghast at this lack of cooperation between the departments less than a month before Vicksburg fell. "The whole situation was treated with a levity incomprehensible when the vast stake is considered," he complained. Kean reported that Secretary of War Seddon "thought there was more blame on the command on the west than on the east side of the [Mississippi] river for [Vicksburg's] loss."[36] As predicted by the bureaucratic model's assumption that "politics does not stop once a decision is made," actors who do not come out of the bargaining process with the result they desired can be expected to be less than enthusiastic about implementation.

During the Vicksburg Campaign, the Confederate departmental system's emphasis on the function of defending autonomous departments inhibited teamwork and unity of effort. Thus Holmes and Taylor focused on their positional functions to secure the Trans-Mississippi and Western Louisiana, respectively, rather than seeing themselves as a part of the Confederate Army team. Thus the fact that their department was secure was more important to them than the fact that the Confederacy was secure. In Kean's estimation, "the fatal notion of making each military Department a separate nation for military purposes without subordination, co-operation, or concert ... lost us Mississippi."[37]

Excessively bureaucratic structures often result in "stovepipes" in which the various departments act independently of each other and tend to further their own interests rather than larger organizational ones.[38] Decision-makers who receive inputs from such a system will often get recommendations that promote a mix of national and bureaucratic interests.[39] Some like Holmes and Taylor will deliberately narrow their focus, but even well-intentioned advisors cannot help but be influenced by their position. To this end, Connelly considers Lee to be an example of an advisor who may have "unconsciously" let his pursuit of his own local interests guide his recommendations about those of the Confederacy writ large.[40] Whether by design or the human condition, the Confederate departmental system highlights that Allison was right. In the bureaucratic model, "where you stand depends on where you sit."

Example 2: The Army–Air Force Helicopter Rivalry, 1950s–1960s

One of the more common manifestations of the bureaucratic model is the competition among the various military services.[41] This interservice rivalry was very apparent in the battle for proponency of the helicopter between the Army and Air Force that emerged after the Korean War. Both services made half-hearted attempts to resolve their differences, such as launching an ill-fated joint testing program, but this and other efforts were largely characterized by competition rather than cooperation. One example of this rivalry was an exchange in the summer of 1964 between General Curtis LeMay, Air Force

Chief of Staff, and General Harold Johnson, his Army counterpart. In response to the Army's use of armed Hueys in Vietnam, LeMay challenged Johnson to an aerial duel. Pulling a cigar from his mouth and gesticulating wildly, he screamed, "Johnson, you fly one of these damned Hueys and I'll fly an F–105, and we'll see who survives. I'll shoot you down and scatter your peashooter all over the ground." This episode can be seen as a microcosm of the overall situation. Based on the "where you stand depends on where you sit" phenomenon, the new helicopter concept was "generally supported by the Army but opposed at every turn by the Air Force."[42]

The debate began as the helicopter emerged as one of the major technological advances in the Korean War. In July 1950, Army units began requesting occasional support from the 3rd Air Rescue Squadron of the Fifth Air Force to evacuate critically wounded soldiers from forward aid stations. The Air Force's Sikorsky H-5s proved highly adept at this task, and the Army steadily increased its requests for such assistance. Recognizing this life-saving capability, the Army formally adopted helicopters for medical evacuation. The Army's 2nd Helicopter Detachment arrived in Korea on November 22, 1950, as the first such unit.[43] The Army intended its helicopters to relieve the Air Force of primary responsibility for battlefield casualty evacuation. The result was that "infighting between the Army and Air Force developed."[44]

After Korea, the Army continued to refine its helicopter operations, emphasizing the air cavalry role. Among the early champions of the air assault concept was General James Gavin, who commanded the 82nd Airborne Division during World War II. He wrote a landmark article in 1954 analyzing the inability of the Eighth Army to exploit the return to maneuver warfare engendered by the Inchon landing in Korea. He concluded that the type of forces needed to conduct long-range reconnaissance, rapid advance, and bypass of obstacles did not exist. In the aptly titled "Cavalry, and I Don't Mean Horses!" Gavin asked,

> Where [were] helicopters and light aircraft to lift soldiers armed with automatic weapons and handcarried light antitank weapons, and also lightweight reconnaissance vehicles, mounting antitank weapons the equal of or better than the Russian T–34s...? If ever in the history of our Armed Forces there was a need for the cavalry arm — airlifted in light planes, helicopters, and assault-type aircraft — this was it.[45]

Pursuant to such a vision, the Army began to use the smaller, turbine-driven UH–1 Huey to supplement strong ground force maneuver by mechanized and armored units. Such Army expansion into air operations created tension with the Air Force, which thought the issue had been settled in a 1948 agreement between Secretary of Defense James Forrestal and the Joint Chiefs of Staff which assigned the Air Force responsibility for all air support, including close air support, reconnaissance, air superiority, logistical air support, support of airborne operations, and interdiction of enemy land power and communications.[46] However, the Army's expansion of organic aircraft and increasing use of the helicopter continued to cause problems between the two services. The result was an example of veteran negotiator Richard Holbrooke's observation that "people sit in a room, they don't air their real differences, a false and sloppy consensus papers over those underlying differences, and they go back to their offices and continue to work at cross-purposes, even actually undermining each other."[47]

Additional attempts to set boundaries followed, such as a memorandum of understanding signed by Secretary of the Army Frank Pace and Secretary of the Air Force

Thomas Finletter in October 1951. This document specified that organic Army aircraft would be used by the Army "as an integral part of its components for the purpose of expediency and improving ground combat and logistics procedures within the combat zone." While matters relating to what functions Army organic aircraft might perform under the exclusive control of the ground force commander were spelled out, the larger issue of roles and missions remained unresolved.[48]

Throughout the 1950s and 1960s, the Air Force sought to maintain flexibility and cost effectiveness by developing multirole, high-performance aircraft that could gain air superiority and then shift between the different tactical air support missions. However, the Air Force's acquisition of such jet multirole fighter-bombers left the Army dissatisfied with what it perceived to be a lack of Air Force commitment to close air support.[49] As a result, the Army appeared to become intent on forming its own air arm, and growth in this direction was rapid. In 1950, the Army inventory included 668 light fixed-wing and fifty-seven rotary-wing aircraft. By 1960, it had over 5,000 aircraft of fifteen varieties. The Army, not the Air Force, was becoming the acknowledged leader in vertical flight and ground-effects assets. The Air Force had even been unable to prevent the Army from "borrowing" three Air Force T-37 jets for testing, and a number of Army aviators were being qualified in various transonic aircraft from other services and the North Atlantic Treaty Organization. As the Air Force found itself with a growing number of "technologically unemployed" pilots, a Congressional committee heard testimony derisively describing the Air Force as the "silent silo-sitters-of-the-seventies."[50]

Rotary-wing aircraft represented different possibilities to the two services. To the Army, helicopters offered the Army a credible means of increasing air support. To the Air Force, they meant great pressure to enhance ground support capabilities or risk losing that mission and the attendant budget to the Army. Within the Department of Defense, the Air Force had reason to worry. In early 1962, Secretary of Defense Robert McNamara reviewed the Army's aviation procurement plan and found it overly cautious. On 19 April, he sent a memorandum to Secretary of the Army Elvis J. Stahr, Jr., saying, "I have not been satisfied with Army program submissions for tactical mobility. I do not believe that the Army has fully explored ... technology for making a revolutionary break with traditional surface mobility."[51]

Because of this failure, McNamara directed that the "reexamination of [Army] aviation requirements should be a bold 'new look' at land warfare mobility. It should be conducted in an atmosphere divorced from traditional viewpoints and past policies." McNamara stated his expectation and stifled bureaucratic naysayers, concluding that he would be disappointed if the "reexamination merely produces logistics-oriented recommendations to procure more of the same, rather than a plan for implementing fresh and perhaps unorthodox concepts which will give us a significant increase in mobility."[52]

Among those McNamara thought capable of the necessary grand vision was Lieutenant General Hamilton Howze, director of Army aviation, and within a week of the Secretary's memo, Howze was appointed president of the Army Tactical Mobility Requirements Board (commonly called the Howze Board). As one author noted, "Seldom has there ever been such a broad and openended charter in military history,"[53] and Howze called the Secretary's memo the "best directive ever written."[54] Howze would take full advantage of the strong mandate presented him. As such, he represents one of many examples of the bureaucratic model's assumption that "policy-making is a social process."

A variety of individuals and institutional actors such as Howze played a key role in the helicopter decision-making process.

After just ninety days, the Howze Board recommended that five Reorganization Objective Army Divisions (ROADs) be replaced by airmobile and air cavalry units. Howze saw the advantage of airmobile forces as mobility, utility in delay operations, ability to ambush, and direct firepower capability.[55] A month after the board reported, the Army deployed fifteen Hueys to Vietnam with a concept team to evaluate their effectiveness in counterinsurgency operations. This team was in addition to the Army's Utility Transport Company which had already been experimenting with the use of armed Huey gunships to escort troop transports and provide suppressive fires on landing zones as early as the autumn of 1962. This existence of an independent Army close air support capability in Vietnam generated consternation among Air Force officers who argued it represented an inefficient duplication of effort and a deviation from their doctrinal tenet that all air power assets be centralized under Air Force control.[56]

Such logic led General Rollen Anthis, commander of the 2nd Air Division, to call for all Army helicopters to be placed under Air Force control. Anthis altruistically claimed he sought only to provide better support for airmobile operations and to increase coordination between Army helicopters and Air Force fixed-wing fighter-bombers. The Army countered that Anthis's proposal would negate whatever advantages the Army had accrued by having organic aviation assets. While Military Advisory Command, Vietnam commander General Paul Harkins did not uphold Anthis's request, in August 1962, Harkins did direct all helicopter assaults be escorted by fixed-wing aircraft and that all enemy positions be suppressed by a concentrated air attack before any landing took place.[57]

As impressive as were the Howze Board's recommendations, the Army decided not to accept them wholesale, fearing they represented too radical an overhaul of the Army's force structure.[58] Instead, the Army leadership ordered more extensive tests, and the 11th Air Assault Division was activated in February 1963 at Fort Benning, Georgia, to fulfill this purpose. The venerable Major General Harry Kinnard, who had served with the 101st Airborne Division during World War II, was selected as the commander. The initial Air Assault I test held at Fort Stewart, Georgia, yielded encouraging results, and a second test (Air Assault II), demonstrated that the "advantages of increased mobility and maneuverability inherent to the air assault division offers a potential combat effectiveness that can be decisive in tactical operations."[59]

In a passionate response to the Howze Board, the Air Force had created its own board chaired by Lieutenant General Gabriel Disosway whose findings not surprisingly refuted the Army's. In contrast to the airmobility concept, the Air Force suggested a joint service combat team structure. Central to the Air Force concept was an assumption that in a joint force, ROAD — supported by Air Force tactical air — offered more practical and economical means of enhancing the mobility and combat effectiveness of Army units than Army air assault divisions. The Air Force proposed that the selective tailoring of ROAD could permit varying degrees of air transportability and combat capability, from a relatively light mobile force to one capable of sustained combat. According to the Air Force, this could be accomplished without specialized airmobile units. Neither Army fixed-wing aircraft nor medium helicopters would be required for tactical movement of troops or resupply because C–130s could accomplish most transport missions while other Air Force aircraft provided reconnaissance and firepower.[60]

A US Air Force Sikorksy CH-3C helicopter transports an Army jeep and its 106 mm recoilless rifle as part of Exercise Goldfire I in 1964 (US Air Force).

While the Army was conducting its Air Assault tests, the Air Force conducted its own parallel Indian River tests at Eglin Air Force Base, Florida.[61] Army troops including an infantry brigade from the 1st Infantry Division at Fort Riley, Kansas, participated in these tests, which were designed in part to allow "the Air Force to refine its mobility concepts through test data before the coming competition with the Army."[62]

The culmination of the Army and Air Force tests was the Goldfire I exercise conducted in October and November 1964, but it was quickly evident that nothing new was being offered with regard to close air support of ground forces. The concept merely streamlined existing procedures and demonstrated that, given heavy dedicated tactical air support, an Army division had increased firepower.[63] After evaluating both the Army and Air Force concepts, General Johnson tactfully summed up his service's dissatisfaction: "I had the rare privilege of seeing the 11th Air Assault one week and the other concept at the early part of the following week, and I would make a comparison of perhaps a gazelle and an elephant. The two are not comparable."[64]

The uninspiring results of Goldfire I and the success of Army tests led in January 1965 to a recommendation by the Joint Chiefs, with the Air Force dissenting, to cancel Goldfire II. Secretary McNamara approved the cancellation, and the Joint Chiefs responded, again with the Air Force in dissent, by recommending approval of the Army request for an airmobile division. In June 1965, McNamara authorized the organization of the 1st Cavalry Division (Airmobile).[65] It was activated in July 1965 and was made up of resources from the 11th Air Assault and the 2nd Infantry Divisions. The division's advance party arrived in Vietnam in late August of that year.

In addition to helicopters, Army aviation plans had included the OV-1 Mohawk, a 12,000-pound plane originally developed by the Marine Corps. The original Bradley-Vandenberg agreement of May 20, 1949, had placed a limit of 2,500 pounds on Army fixed-wing aircraft.[66] Technological developments led the Army to press for a relaxation of this restriction, and an agreement between Pace and Finletter of November 4, 1952 raised the limit to 5,000 pounds.[67] Nonetheless, the Army successfully appealed to Secretary of Defense Charles Wilson that if the Marine Corps, despite having jets, also needed the capabilities of the OV-1, the Army probably did too.[68] It addition to the observation function of the OV, an AV version was also used as an attack aircraft to deliver ordnance in a close air support role.[69]

The Howze Board had also concluded that the Air Force C-130 transports would be unable to operate from the rough forward airstrips that would be necessary in airmobile operations.[70] The Army's solution was the CV-2 Caribou, a modification of an off-the-shelf de Haviland twin-engine, high-wing commercial cargo transport designed to operate in the northern Canadian wilderness. Although its carrying capacity was less than a third of the C-130's, the Army considered the Caribou's short-field operating ability decisive.[71]

The program recommended by the Howze Board included 395 fire-support Mohawks and 400 Caribou transports. The idea of the Mohawk as a close-support aircraft, particularly one that was organic to an Army division, was particularly disconcerting to the Air Force.[72] Although both these planes represented the Army's optimal solution, the Air Force saw them as dangerous threats to traditional roles and missions and the Air Force's doctrine of centralized control of aircraft.[73] The bureaucratic model assumes "policy decisions are compromises," and in this case compromise was indeed required. As "a sacrifice on the altar of accord with the Air Force," General Johnson was forced to withdraw Army plans to use the Mohawk as an attack aircraft, confining it to reconnaissance. Later, he was also compelled to relinquish the CV-2s, which the Air Force redesignated C-7s.[74]

What Johnson and the Army got in exchange for the Caribou was that the agreement "essentially established without a doubt the Army's claim to the helicopter, and especially the armed helicopter." Securing this ability was absolutely critical. As Lieutenant General John Tolson explains, "the keystone to airmobility was—and is—the helicopter." It "was the absolute *sine qua non* of the Army's concept of mobility." Johnson, Tolson writes, "was keenly aware of this basic fact."[75] In exchange for this accommodation, the Army was able to keep the armed Huey, the most critical component of its new airmobility concept.[76] Most of the Army chain of command was accepting of the compromise, though at the same time lamenting what might have been. "The Army-Air Force trade-off cost us the Caribou," noted Major General Ellis Williamson who commanded the 173rd Airborne Brigade in Vietnam. "On the balance it was good. General Johnson, in my opinion, made the decision he should have. But Army aviation would have gone considerably further if he hadn't had to make that decision." In the end, Williamson felt Johnson "got the most he could under the circumstances."[77]

Because the participants surrendered some of their preferences in the compromise, the bureaucratic model also assumes "politics does not stop once a decision is made."

While the 1966 agreement between Johnson and his Air Force counterpart General John McConnell may have resolved Army's Air Force tactical air transport competition by eliminating the Army's fixed-wing combat aircraft, it paradoxically also created close-air support competition by officially sanctioning a role in this area for both services. Soon the Army and Air Force began disputing the Army's plans to develop the AH-56 Cheyenne as an attack helicopter.[78] Part of the Air Force's objection was that the Cheyenne had "stub wings," as well as its rotors, and was therefore not a helicopter at all.[79] This debate was resolved in the Army's favor by the Office of the Secretary of Defense in 1968, and, although the AH-56 program was abandoned in 1969 due to technical problems, the replacement AH-64 Apache program was successful.[80]

Although joint operations have advanced dramatically since Vietnam, tension between the Army and Air Force over close air support remains and periodically resurfaces. For example, as the Army sought to reengineer itself to deal with the post–Cold War environment, Douglas Macgregor noted one consideration was that Army "reconnaissance and attack helicopters have been developed to acquire permanently a close air support capability that receives low priority in the US Air Force."[81] He predicted the trend would continue into the twenty-first century, writing, "Modern air defense systems will drive jet-driven aircraft to higher and higher altitudes with the result that stealthy, rotor-driven aircraft along with unmanned strike aircraft will gradually supplant traditional airframes in the close air support role."[82] Indeed, as the US waged its post–September 11 long war against terror, controversy reemerged between the Army and Air Force over control of unmanned aerial vehicles (UAVs). Both services agreed that the Army should control tactical level UAVs like the Raven, and the Air Force should control strategic UAVs like the Global Hawk. The grey area revolves around control of the extended-range multi-purpose vehicles such as the Predator, that fall between the strategic and the tactical.[83] In another controversy reminiscent of the Caribou debate, the Army and Air Force have argued over a joint heavy lift helicopter that would deliver large cargoes into areas without developed landing strips.[84]

Perhaps the continuing Army's Air Force debate over roles and missions can be explained by the bureaucratic model's assumption that "there is no single version of the national interest." While both services are undoubtedly committed to the defense of the United States, they also have their own parochial perspectives and interests. Ian Horwood is a very pointed in his criticism of this aspect of bureaucratic competition, arguing that little has changed between the services since the 1960s helicopter controversy. "There is no reason to believe," Horwood asserts, "that each service has discarded the sense of its own preeminence that contributed so much to interservice dispute over airpower issues during the Vietnam period."[85] Certainly Horwood's conclusion points to one of the bureaucratic model's flaws in producing what may be the suboptimal decision for the larger interest.

Example 3: The Department of State and the Department of Defense and the Multinational Peacekeeping Force in Beirut, 1982

In the summer of 1982, outside interference and internal political competition among Lebanon's religious groups had plunged the country into its latest in a long series of secu-

US Special Envoy Philip Habib, shown here on December 1, 1982, in Beirut wearing the grey jacket (center), negotiated the agreement that resulted in the deployment of the Multinational Force. Secretaries Shultz and Weinberger held markedly different views about the virtues of Habib's arrangement (photograph by R.P. Fitzgerald; U.S. Navy).

rity crises. Hoping to ease the tension, American Special Envoy Philip Habib arranged for a withdrawal of both Palestinian Liberation Organization (PLO) and Israeli forces from Beirut and the introduction of a peacekeeping force called the Multinational Force (MNF) that originally included US, Italian, and French contingents.[86] Elements of the 32nd Marine Amphibious Unit (MAU) commanded by Colonel James Mead had landed in Lebanon on June 23 to evacuate American citizens from Juniyah. As a result of Habib's negotiation, Mead's men now became part of the MNF.[87] This first MNF remained in place until September 10, when, considering its mission accomplished, it was disbanded and departed Lebanon.

The departure of the MNF left Beirut with no functioning civil authority, and the respite was not to last long.[88] Four days after the MNF's departure, Bashir Gemayel, the Maronite Christian chief of the Phalangist militia and president-elect of Lebanon was assassinated. Within days the Israeli Army moved back into Beirut, and the Phalangist militia, now led by Elie Hobeika, staged a massacre of some 700 unarmed Palestinians in the Sabra and Shatila refugee camps. The Israeli Army, commanded by Brigadier General Amos Yaron, remained on the periphery during the atrocities.[89] Ambassador Habib and President Gemayel had both promised that the Palestinian refugees would be safe after the departure of the PLO from Beirut, but now it appeared that "the Israelis were going in for the kill."[90] Fearing further instability, Bashir's brother Amin Gemayel, who had been elected as the new Lebanese president, appealed for American and European help. On September 20, President Ronald Reagan announced the US, France, and Italy would reconstitute an MNF to replace the Israeli Army and the Phalangists in Lebanon until the Lebanese Army could assume that responsibility. On September 29, Colonel Mead's redesignated 22nd MAU returned to Beirut.[91]

Mead's second stay in Lebanon was short. On October 30, the 22nd MAU was relieved by the 24th MAU, commanded by Colonel Thomas Stokes. The 24th MAU remained until February 15, 1983, when Mead returned for a third time with the 22nd MAU. On May 30, Colonel Tim Geraghty and the 24th MAU assumed the mission again. Increasing violence against the MNF culminated on October 23 when a terrorist attack on the marines' headquarters and barracks left 241 US servicemen dead. The vulnerable position Geraghty's command found itself in can in part be explained by the bureaucratic decision-making model that brought the MNF to Lebanon.

There has historically been a tension between the Department of State and the Department of Defense in most US presidential administrations, but the struggle between Secretary of Defense Caspar Weinberger and Secretary of State George Shultz was particularly heated.[92] Shultz had replaced Alexander Haig on June 25, 1982, almost simultaneously with the deployment of the 32nd MAU. At the time, a hopeful Andrew Knight concluded that Shultz's arrival had even contributed "in the Middle East to the start of a possible peace process."[93] Knight was also cautiously optimistic that Shultz might bring some relief in the "battle between Defense and State and their respective allies in other quarters" that had left President Reagan "the third point in a triangle" between Weinberger and Haig.[94] The tension between the two secretaries had been so personal that "close associates of both reported at an early stage that there was an almost animal dislike between the two."[95] On the other hand, Shultz and Weinberger held similar policy views on the Middle East, and observers took heart in finding "a President and his two principal Secretaries in rare accord."[96]

But if Shultz and Weinberger had similar geopolitical views on the Middle East, they had different philosophies on the use of force. As Secretary of State, Shultz "stressed the need to interweave diplomacy and force in a wide range of situations."[97] While he argued for the "prudent, limited, proportionate uses of our military power,"[98] he complained that "to Weinberger, as I heard him, our forces were to be constantly built up but not used: everything in our defense structure seemed geared exclusively to deter World War III against the Soviets; diplomacy was to solve all the other problems we faced around the world."[99]

For his part, Weinberger interpreted the State Department's position as being that

> we should not hesitate to put a battalion or so of American forces in various places in the world where we desired to achieve particular objectives of stability, or changes of government, or support of governments or whatever else. Their feeling seemed to be that an American troop presence would add a desirable bit of pressure and leverage to diplomatic efforts, and that we should be willing to do that freely and virtually without hesitation.[100]

Weinberger considered the National Security Staff to be "even more militant" with an "eagerness to get us into a fight somewhere — anywhere — coupled with their apparent lack of concern for the safety of our troops, and with no responsibility therefor."[101] He was particularly troubled by this development because rather than serving as a mediator and honest broker in the debates between the State and Defense Departments, Robert "Bud" McFarlane, first as the deputy and then as the actual National Security Advisor, often sided with Shultz in policy disputes.[102] Indeed, in the bureaucratic model, it is not uncommon for actors to negotiate and bargain internally.[103] This time, bureaucratic forces aligned themselves at the expense of the Department of Defense.

For his part, Weinberger believed that "we should not commit American troops to any situation unless the objectives were so important to American interests that we had to fight." Even then, the commitment of the military would be "as a last resort" to be used only after "all diplomatic efforts failed."[104] Because Weinberger did not think these conditions had been met in Lebanon, he opposed the deployment of the second MNF. The Joint Chiefs of Staff shared Weinberger's opposition to the plan, but the State Department and McFarlane joined forces and continued to strongly press for a second MNF. In the end, what Weinberger calls McFarlane's "petulant" demands carried the day.[105] Out of loyalty and professionalism, Weinberger "supported the President's decision fully," but Thomas Hammes argues there certainly was "no attempt to seek the effective interagency response essential to bring peace."[106]

While Weinberger maintained the MNF was more of a vulnerability than a threat,[107] Shultz pursued his theory that diplomacy could work "most effectively when force — or the threat of force — was a credible part of the equation."[108] Eight months later, Shultz's efforts produced what Weinberger called "a curious agreement" between Lebanon and Israel. "Why such an agreement was reported to us in such glowing terms by George Shultz," Weinberger said, "has always remained a mystery to me." Weinberger raised his objections with Shultz before the agreement was signed, but Shultz remained "extremely proud and protective of his agreement, and none of [Weinberger's] arguments ... made the slightest impression on him."[109]

Such parochialism is detrimental to a decision-making process theoretically designed to serve the national interest. Instead, as Eugene Wittkopf and his colleagues note, "the president is surrounded on a daily basis by advisers, including members of the cabinet, who so interpret their jobs as to make maximum claims on their agency's behalf."[110] Thus the MNF decision-making process reflects the bureaucratic model's assumption that "there is no single version of the national interest," or in Allison's words, "where you stand depends on where you sit."

Another bureaucratic model assumption borne out in this case study is the idea that "policy-making is a social process." As such, decisions are made in an abstract political space which may have little relevance to the practical requirements or situation of those who will actually have to implement the decision. These "boots-on-the-ground" interests are often better represented by the involvement of a hierarchical chain of command. Lieutenant Colonel H. L. Gerlach, who commanded the Battalion Landing Team (BLT) 1/8, the ground element of the 24th MAU when it assumed the MNF mission on May 30, 1983, explained, "The political and diplomatic side of the house set up the parameters, and we accomplish our mission within them."[111] These parameters, however, were critical, and Weinberger insisted that the military's success "was premised on achieving a *diplomatic* success."[112] Colonel Tim Geraghty, commander of the 24th MAU, would come to believe the failure to establish this diplomatic success was his unit's ultimate undoing. "When you have the State Department leading the way, and the politicians fixing our bayonets," he criticized, "it becomes a loser's game."[113]

While Secretary of State Shultz saw the second MNF as a tool to facilitate his ongoing negotiations, Secretary of Defense Weinberger objected that it did "not have any mission that could be defined. Its objectives were stated in the fuzziest possible terms."[114] President Reagan wanted to communicate America's strategic interest in Lebanon, but it was unclear how this interest translated to an operational mission. In his September 29, 1982, message to Congress, Reagan said the MNF's "mission is to provide an interposition force at agreed locations and thereby provide the multinational presence requested by the Lebanese government to assist it and the Lebanese Armed Forces." [115] The September 23 Joint Chiefs of Staff Alert Order used similar language, describing the US forces "as part of a multinational force presence in the Beirut area to occupy and secure positions along a designated section of the line south of the Beirut International Airport to a position in the vicinity of the Presidential Palace." [116] It was clear where the MNF was supposed to be, but what it was supposed to do there and why remained vague.

What Colonel Geraghty concluded was that "the mission of the MAU in Lebanon is a diplomatic mission." "It was important to me, in the interpretation of that mission," he explained, "that there was a presence mission. That means being seen." [117] According to Robert Jordan, the Marine Corps public affairs officer in Beirut and later executive director of the Beirut Veterans of America, "Presence was interpreted to mean a showing of the flag, a symbol of American interest and concern for the legitimate government of Lebanon and a neutral stance toward Israel, Syria, and the various religious and political factions." [118] Later, the "mission" was defined to be the interposition of the MNF between the withdrawing Israelis and Syrians, but the problem remained, of course, that there was not an agreed-upon withdrawal. [119] "Absent this," Weinberger writes, "there was no *military* action that could succeed, unless we declared war and tried to force the occupying troops out of Lebanon." He argues that even after the "objective was 'clarified,' the newly defined objective was demonstrably unobtainable." [120] As the situation declined, the MNF lost its legitimacy with the Muslim population and the violent militias that represented it. Unaware of the ramifications of this change, the MNF continued to conduct itself with restraint based on its understanding of being a peacekeeping force and failed to improve its security posture.

The MNF case study also illustrates the bureaucratic model's assumption that "politics does not stop once a decision is made." Shultz certainly felt the Department of Defense acted according to this premise, stating, "What I heard [Weinberger] advocate gave me my first experience with what I would come to recognize as a standard Pentagon tactic: when you don't want to do something agree to do it — but with such an impossible set of conditions and on such a preposterously gigantic scale that the outcome will be to do nothing." [121] Although he was forced out of deference to the president to acquiesce to Reagan's desire to deploy the MNF, Weinberger's insistence on the condition that all foreign forces had first agreed to a withdrawal and had in fact departed Lebanon amounted to what Shultz would characterize as the bureaucratic model's tendency to result in half-hearted or uncooperative implementation of a decision that did not reflect the agency's interests.

Perhaps the most damning aspect of the MNF decision is its demonstration that in the bureaucratic model, "policy decisions are compromises." Almost by definition, compromises of this sort result in satisfying rather than optimizing policies. [122] In the case of the MNF, neither side got what it wanted nor what it thought was necessary to result in complete success. Instead, as Shultz describes it, "the military wants to do what the diplomats don't think is necessary, and the diplomats want the military to do what the military is too nervous to do." [123] The result was an untenable position for the MNF.

If Shultz's philosophy of using diplomacy and force simultaneously had carried the day in the decision to deploy the MNF, the terrorist attack of October 1983 seemed ultimately to have validated Weinberger's counter argument. He codified his lessons learned in the "Weinberger Doctrine"—strategic criteria to help guide "the painful decision that the use of military force is necessary to protect our interests or to carry out our national policy."[124] The Weinberger Doctrine proved a useful and unifying tool in such scenarios for the duration of the Cold War, but afterwards the competition between the Departments of State and Defense resurfaced as parts of decisions to intervene in places like Somalia, Bosnia, and Kosovo, showing the bureaucratic model remains a fixture in American policy-making concerning the use of force.

Utility of the Bureaucratic Model

The rational actor model depicts the decision-making process as being the work of unitary actors pursuing choices that maximize national interests. The model's critics argue this assumption is at best a great oversimplification. Instead, they see decisions as resulting from a struggle for influence and power driven by differences of interest. According to the bureaucratic model, decision-making is a social and political process rather than the intellectual and administrative one envisioned by the rational actor model.[125] The bureaucratic model also has advantages over poliheuristic theory in analyzing group decision-making processes rather than those involving a single individual.

By focusing on internal politics, the bureaucratic model views decisions as "outcomes of various overlapping bargaining games among players arranged hierarchically in the national government." While the players in the bureaucratic model have their own "perceptions, motivations, positions, power, and maneuvers," they are still parts of larger organizations with their own standards patterns of behavior. The organizational process model recognizes this reality and views decisions not as the *choices* of the rational actor model or the *outcomes* of the bureaucratic model, but as "the *outputs* of large organizations functioning according to regular patterns of behavior."[126] Indeed, there are countless cases where decisions are made routinely and rapidly on the basis of some *a priori* guideline or administrative rule. In such instances, the organizational process model may be a greater analytical tool.[127]

5

The Organizational
Process Model

Over time, organizations find themselves repeatedly dealing with similar situations and performing recurring tasks. When they discover a comfortable and successful way of dealing with these frequent occurrences, they develop standard operating procedures that promote efficiency and reduce uncertainty in complex situations.[1] The assumption behind these standard procedures is that performance averaged over a range of cases will be superior than it would be if each case were approached individually.[2] Thus, instead of formulating a reasoned and specific response to a new situation, the organizational process model argues that decision-makers act based on how the new situation corresponds to behavior learned in previous situations.

The organizational process model manifests itself in a variety of ways including a reliance on a frame of reference that provides context to the new situation, incrementalism that protects the organization from catastrophe, standard operating procedures that save time, and change that occurs after a failure. The formulation of American strategy during the Vietnam War, the decision to abort the Iranian hostage rescue mission in 1980, and the federalization of the California Army National Guard during the Los Angeles Riots in 1992 all illustrate the organizational process model.

Example 1: Strategic Formulation
in the Vietnam War, 1960s–1970s

The US prosecution of the Vietnam War is rife with examples of the organizational process model. Unfortunately from the American perspective, its application largely produced disappointing results. The effort to restructure the South Vietnamese Army, the Strategic Hamlet Program, the conventional strategy of attrition, and the graduated nature of the bombing campaign all showed the dangers of an over-reliance on previous experience and familiar procedures in new and different situations.

The American inclination to use an existing model in a new situation surfaced as early as the advisory period — even before US ground troop involvement in Vietnam. According to Tran Van Don, an officer who participated in the first planning sessions in 1954 to reorganize the South Vietnamese Army, the South Vietnamese favored patterning their army after that of the Viet Minh. However, Lieutenant General John O'Daniel, the first chief of the American advisory effort, insisted on a US style organization in order to facilitate American logistical support. The result was "a nifty miniature copy of the US military establishment."[3] This early emphasis on "Americanization" did not prove beneficial to the later requirement for "Vietnamization" and illustrates the dangers of assuming that what works for the United States will also work for the nation the United States is assisting. A RAND Note entitled "Countering Covert Aggression" would indicate that this is not a tendency unique to the American experience in Vietnam. The report notes the US propensity

> to shape Third World forces too closely in the image of US forces, which have not been designed for counterinsurgency warfare; to provide Third World forces with high-technology weapons and equipment that are too costly for and ill-suited to local capabilities and likely battlefield requirements; and to make Third World forces overly reliant on US advice and support, a situation that could seriously impair local capabilities in the event of US aid cutbacks.[4]

Another example of reliance on a previous model was the Strategic Hamlet Program. Of all the situations that made the South Vietnamese rural population vulnerable to Viet Cong (VC) exploitation, perhaps the most frustrating was the critical need for land reform. Based on a program that had worked well for the British in Malaya, the Strategic Hamlet Program was designed to concentrate the rural population in a limited number of fortified villages to provide them physical security against the VC. Once security was established, social programs that would hopefully foster government allegiance were planned to follow.[5]

The Strategic Hamlet Program was largely a failure because it simply did not recognize the realities of the new situation. Unlike the Chinese immigrant squatters who were the subject of the British relocations in Malaya, the Buddhist South Vietnamese had ancestral ties to the land, and moving interrupted their practice of veneration of ancestors. Additionally, the relocations caused the peasants to abandon generations of hard work and took vital, arable land out of production, which hampered economic progress. In the new hamlets, the peasants had to start over from scratch, without compensation for their labor or loss. These factors obviously led to a disgruntled population that was ripe for VC exploitation — a situation facilitated by the fact that many VC secretly relocated to the new hamlets with the rest of the population. Many peasants were so alienated by the entire ordeal that they slipped away from the hamlets and returned to their ancestral lands. This development greatly hindered one of the goals of the relocation, which was to create free fire zones in the vacated areas based on the assumption that anyone there now was a VC.[6]

The relocations created other problems as well, including the perception that if relocation was necessary in the first place, then security must be weak. Many peasants were left with the impression that if the South Vietnamese government was not able to secure even its allies, fully supporting the government would be dangerous. Finally, by moving the population away from the countryside, a significant, if imperfect, source of intelligence was lost.[7] Summing up the Strategic Hamlet Program's failure to address the needs of the South Vietnamese people, Lieutenant General Dave Palmer concludes the program was "executed with too little real feeling for the human beings involved."[8] Part of that problem was thinking the perspectives of the recent immigrants in Malaya were the same as those of the native Buddhists in South Vietnam.

Andrew Krepinevich argues that the strategic vision of Military Advisory Com-

President Kennedy turned to his special military advisor, former Army Chief of Staff Maxwell Taylor, in an effort redirect the Army's strategic thinking (photograph by Abbie Rowe; John F. Kennedy Presidential Library and Museum).

mand, Vietnam commander General William Westmoreland was another example of this pattern of relying on a previous experience of limited applicability to the present situation. Inspired by the formula that had proved successful in World War II, Westmoreland's concept for Vietnam focused on "conventional war and reliance on high volumes of firepower to minimize [US] casualties." "Unfortunately," Krepinevich notes, "the Army's experience in war did not prepare it well for counterinsurgency where the emphasis is on light infantry formations, not heavy divisions; on firepower restraint, not its widespread application; on the resolution of political and social problems within the nations targeted by insurgents, not closing with and destroying the insurgent's field forces."[9] The result according to Jeffrey Record was that "Army leaders looked upon irregulars with disdain and believed that conventional forces that had defeated German armies could readily handle a bunch of rag-tag Asian guerrillas."[10] Based on this understanding, Westmoreland pursued a strategy of attrition that used battalion-size and larger "search-and-destroy" operations to "find, fix, flush, and finish" the enemy in a way that maximized traditional American reliance on the offensive, mass, firepower, and technology. Russell Weigley writes that it was a technique "sanctioned by the most deeply rooted historical conceptions of strategy."[11]

There certainly were options to attrition. President John F. Kennedy had noted that guerrilla warfare requires "a whole new kind of strategy, a wholly different kind of force, and therefore a new and wholly different kind of military training."[12] Kennedy turned to Maxwell Taylor to head a high-level interdepartmental "Special Group, Counter-Insurgency" to begin developing the new strategy.[13] With this new emphasis on limited war, the Army was persuaded to drop from its 1962 edition of FM 100–5, *Army Field Service Regulations* the familiar statement that "the ultimate objective of all military operations is the destruction of the enemy's armed forces and his will to fight."[14] But as Jack Levy and William Thompson explain, "instead of looking forward and calculating which strategy would lead to the optimal outcome under the circumstances," the organizational process model "looks backwards to those routines that the organization set up to automatically implement certain types of policies that provide the best fit to the problem at hand."[15] Indeed, in developing its strategy for Vietnam, the Army leadership "preferred to proceed as though the Field Service Regulations had never changed."[16] Graham Allison offers some insight into this ability of the Army to resist change by noting that "government leaders can substantially disturb, but not substantially control, the behavior of [the] organizations" of which a government consists.[17]

The establishment of CORDS (Civil Operations and Revolutionary Development Support) in 1967 offered new hope for proponents of pacification as an alternative to attrition. Ambassador Robert Komer, head of the new organization, knew there could be "no civil progress without constant real security" for the South Vietnamese population.[18] Recognizing this prerequisite, he pragmatically argued, "Until the [South Vietnamese Government] regained dominant control of the countryside and provided credible semipermanent protection to the farmers, it would hardly be feasible to proceed with other aspects of pacification."[19] In the same vein, Maxwell Taylor stated, "We should have learned from our frontier forbears that there is little use planting corn outside the stockade if there are still Indians around the woods outside."[20] Indeed a 1966 Army study called Program for the Pacification and Long-term Development of South Vietnam (PROVN) had urged this same philosophy. Nonetheless, Komer lamented that even at its height, pacification "remained a small tail to the very large conventional military dog."[21]

The program that probably came closest to the intent of PROVN was the Combined Action Program or CAP. Beginning as a small experiment to secure US military bases around Phu Bai and Da Nang in 1965, CAP soon became the linchpin in the Marine Corps' strategy for winning the war. [22] It also came to exemplify Westmoreland's negative attitude toward pacification.

A CAP platoon was a combination of a fourteen-man Marine Corps rifle squad and one Navy medical corpsman, all who were volunteers, and a locally recruited Popular Forces (PF) platoon of about thirty-five men. The resulting CAP was assigned responsibility for a village, which typically consisted of five hamlets spread out over four square kilometers with an average population of 3,500 people. The American Marines lived with their Vietnamese PF counterparts and became integral parts of the unit. The effect was synergistic. The Marines gained intelligence from the South Vietnamese soldiers' knowledge of the local terrain and enemy, while the PF benefited from the Marines' firepower, tactical skills, and discipline. The CAP was a solid and mutually beneficial combination.[23]

Perhaps most important, the constant Marine presence sent a powerful message that the Americans were there to stay. They did not fly in by helicopter in the morning and fly out at night to leave the villagers at the mercy of the VC. This continued presence was critical because the peasant who cooperated with the government had to carefully weigh the risk of VC reprisals against himself, his family, his friends, and his community with the benefits of improved clothing, food, education, and medical assistance. When the Americans flew in and flew out, the risks to the Vietnamese villager often outweighed the benefits. However, under CAP, the Marines shared the same fate as the South Vietnamese soldiers and people. In fact, CAP Marines took two and a half times the casualties of the PF in the CAP. The CAP was a strong testimony of American commitment and partnership, and gave the Vietnamese people a sense of enduring security.[24]

The CAP program expanded steadily, and in 1966 there were fifty-seven CAP platoons. By the end of 1967, the number had grown to seventy-nine. Despite these increases and demonstrated success, Westmoreland was unwilling to adopt the program, arguing that he "simply had not enough numbers to put a squad of Americans in every village and hamlet; that would be fragmenting resources and exposing them to defeat in detail."[25] While there is some merit to Westmoreland's argument about numbers, his genuine objection lay more in a fundamental strategic difference. Westmoreland viewed the CAPs as static and defensive employments of his resources. Instead, he favored the aggressive pursuit and destruction of enemy forces that had served the Army well during World War II. The focus of CAP at the small unit level also violated Westmoreland's quest for the mass he needed to gain a conventional battlefield victory.[26]

Westmoreland never obtained the concentrated overwhelming force he wanted, and one of the most criticized aspects of the American strategy in Vietnam was its "graduated response."[27] Such an approach is an example of the organizational decision dynamic David Braybrooke and Charles Lindblom describe as "incrementalism." This conservative approach consents to fine tuning past decisions but does not include a broad exploration of policy alternatives. By allowing no large deviation from past choices, incrementalism protects against catastrophic failure from any one decision. If left unchecked, however, incrementalism can get out of control. The American experience in Vietnam is considered by many observers to represent "incrementalism run amok."[28]

The incrementalistic approach prompted both Presidents Kennedy and Johnson to make minimally necessary decisions in an effort to avoid a full US intervention. Eventually, Johnson got sucked in.[29] One of the formative moments in the evolution of the graduated response was what Assistant Secretary of Defense John McNaughton described as the "slow squeeze" approach to the bombing campaign. Lieutenant General Palmer writes that the plan was "to start softly and gradually increase the pressure by precise increments which would be unmistakably recognized by Hanoi." According to McLaughton, such a "scenario would be designed to give the United States the option at any point to proceed or not, to escalate or not, and to quicken the pace or not." Although the Joint Chiefs of Staff preferred to apply overwhelming force, President Johnson approved this incremental approach. He also insisted on being very involved in the target selection process, boasting at one point that the Air Force "can't even bomb an outhouse without my approval." Thus in spite of the bombing campaign's ominous name of "Rolling Thunder," it was in execution an incremental exercise in fits and starts.[30]

The slow squeeze strategy included a parallel diplomatic track so bombing pauses were provided for to give opportunities for negotiations.[31] As Admiral Thomas Moorer explains:

This gradual application of airpower, with frequent bombing halts over the course of time, was intended to give the enemy pause and motivate him into seeking a political settlement of the war. Instead, the gradualism actually granted the enemy time to shore up his air defenses, disperse his military targets, and mobilize his labor force for logistical repair and movement. From a military point of view, gradualism violated the principle of mass and surprise which airpower has employed historically to attain its maximum effectiveness. Gradualism forced airpower into an expanded and inconclusive war of attrition."[32]

Rolling Thunder bombing sorties grew from 25,000 in 1965 to 108,000 in 1967. The tonnage of bombs dropped likewise increased from 63,000 to 226,000, exceeding the totals dropped on Germany, Italy, and Japan in World War II. The North Vietnamese, however, countered the gradual increases by perfecting a technique of "fighting while negotiating." Keenly aware of US domestic politics, including election cycles, the communists proved to be shrewd bargainers. For example, when President Johnson made a focused attempt to reach a negotiated settlement prior to the November 1968 elections, the North Vietnamese exploited the situation. On October 31, on the basis of informal, unwritten "understandings"—which the North Vietnamese neither officially accepted nor rejected—the US completely halted its bombing campaign. Having achieved the desired objective, the North Vietnamese then proceeded to simply ignore the "understandings." Later, when President Richard Nixon tried to negotiate through intermediaries in the summer and fall of 1969, the North Vietnamese merely dragged out the negotiations in order to buy time to recover from Tet and to pressure the US to make concessions. In the process, vital resources were protected, damaged infrastructure was prepared, and external support from China and the Soviet Union was increased at rates that more than offset losses. More importantly, the communists' will to keeping fighting was never broken.[33] Indeed, incrementalism's small changes to the status quo usually provide only temporary relief and rarely are able to completely solve problems.[34]

If there is a common theme in these four examples, it is the organizational process model's tendency to "fight the last war." This is not to say that effective decision-makers should not build a personal frame of reference from schooling, experience, self-study, and assessment. Reflecting on past experiences promotes learning and helps place the current situation in strategic context.[35] However, the frame of reference is designed to expand, not limit, the decision-maker's horizons. He cannot unimaginatively apply a course of action that worked once to a new situation for which it is inappropriate. Instead, decision-makers must be mentally agile enough to understand the circumstances around them and adjust. This was a large part of the American failure in Vietnam. Techniques that worked in America, Malaya, or World War II were assumed to be transferrable solutions to Vietnam. As it turned out, this misapplication of frame of reference under the auspices of the organizational process model inhibited, rather than facilitated, the American response to the new environment.

Example 2: The Decision to Abort
the Iranian Hostage Rescue Mission, 1980

Mohammad Reza Pahlavi became the Shah of Persia in 1941. The US helped solidify his hold on power in 1953 when a CIA-directed operation ousted Prime Minister Mohammed Mossadegh who had nationalized the Iranian oil industry and generated fears that Iran might move closer to Moscow. However, while the Shah brought prosperity to Iran, his policy of modernization and secularization gradually put him in conflict with Iran's Shi'a clergy.

The Shah resisted this opposition, quashing a confrontation in 1963 and exiling the movement's leader, Ruhollah Khomeini, to Iraq. He also attempted to discourage further dissent with strong-armed crackdowns by his secret police. Still, the inequitable distribution of wealth in Iran and the Shah's western-leaning attitude fueled a growing opposition that forced him to flee to Egypt on January 16, 1979. Two weeks later, Khomeini returned to Iran to a hero's welcome.

By this time, cancer had left the Shah in a serious medical condition, and on October 23, President Jimmy Carter allowed him to come to New York for treatment. This decision was the ostensible justification for Iranian students to seize the American Embassy in Tehran on November 4. Khomeini backed their action, seeing it as an opportunity to consolidate power. Initially, the Iranians held sixty-six hostages, but they released thirteen women and blacks as a "humanitarian" gesture, leaving fifty-three still captive. While the majority was on the American Embassy compound, three were held at the Iranian Foreign Ministry—a situation which would complicate any rescue attempt.[36]

Policy makers considered options of waiting for the Iranian internal situation to stabilize, diplomacy, economic and political sanctions, mining Iranian harbors, and an all-out military attack, before deciding to attempt a hostage rescue mission.[37] Rose McDermott has used prospect theory effectively to explain the decision to choose this option, but the organizational process model is useful in explaining the decision to abort the mission once it was initiated.[38] Indeed, in the organizational process model, senior leadership has its greatest influence in initially determining the general policy. Once that policy is selected, lower-level actors within the organization will make decisions according to standard procedures. Thus in the case of the Iranian hostage rescue attempt, Gary Slick argues that "once the decision was made to proceed with the mission, [President Carter] left the details in the hands of his military specialists."[39]

Those details were ironed out at a series of planning conferences, including one held at Fort Bragg, North Carolina, on December 2. There planners decided to infiltrate and exfiltrate a Delta Force rescue team by helicopter. The Navy's RH-53D helicopter was later selected for the mission because of its range, capacity, folding rotor blades for aircraft carrier operations, and operational security. An antisubmarine warfare squadron equipped with the RH-53D was attached to the joint task force (JTF), but the senior Marine in the Joint Chiefs of Staff, Lieutenant General Phillip Shulter, recommended that Marine pilots would be better prepared for the low-level, blacked out flying using night vision devices than the Navy pilots would be. Shulter's recommendation was accepted, even though the Marine pilots were not experienced with the Navy's RH-53D helicopter.[40]

The mission was codenamed Operation Eagle Claw, and Major General James Vaught was designated the joint task force commander. Colonel Charlie Beckwith would command the Delta Force operators that would perform the rescue. Colonel Chuck Pittman was the senior member of the Marine pilot contingent, and Lieutenant Colonel Edward Seiffert was designated the flight leader and pilot of the lead helicopter. These men and a select group of other key commanders and staff met again at Fort Bragg on January 4 and 5, 1980, to continue planning. At this meeting, Beckwith and Pittman took up the important subject of the helicopter requirement.[41]

As planning progressed, the size of the force needing helicopter transportation had grown from about eighty to as many as 132. With these increases, the helicopter requirement also grew to eight RH-53Ds. An important part of air movement planning is to establish "abort criteria" which address "a change of one or more conditions that seriously threatens mission success."[42] As Pittman and Beckwith discussed this aspect of the plan, they recalled that their experience in Vietnam had taught them that "if you needed two helicopters, because of their undependability, you actually had to have three." They agreed that the mission could not leave the refuel site codenamed Desert One for the hide site codenamed Desert Two with fewer than six helicopters. The planners went on to establish helicopter abort criteria for the other phases of the mission. Seven RH-53Ds would be required for launch from the USS *Nimitz* and five to begin the actual embassy assault and extraction.[43] These abort criteria would become the standard operating procedure in the context of organizational process decision-making once Eagle Claw was launched.

The plan was for the Delta operators to be transported from the island of Masirah, off the coast of Oman, to the Desert One site in Iran on board three MC-130 airplanes, accompanied by three fuel-bearing EC-130s. There they would rendezvous with the eight RH-53D helicopters launched from the *Nimitz* somewhere in the Gulf of Oman. The helicopters would fly on different routes in four sections of two each, arriving approximately thirty minutes after the last plane had landed. They would refuel and then on-load the assault force for transport to Desert Two.[44]

On April 24, the eight helicopters took off from the deck of the *Nimitz* headed for Desert One according to the plan. Some two hundred miles out, a warning light in one of the helicopters indicated a problem with the pressurization in a rotor blade. For the Marine pilot who was used to flying CH-53s, such an indicator meant that an imminent crash was possible, and the helicopter should be landed. What he did not know because of his unfamiliarity with the more advanced Navy helicopter he was now flying, was that a RH-53D had never crashed after such an indication.[45] As John Valliere notes, "When the Marines were signed on for the JTF, so was their experience."[46] The organizational process model would expect the pilots to act in accordance with "prior experience in dealing with like situations."[47] Indeed, based on his CH-53 frame of reference, the Marine pilot abandoned his helicopter in Iranian territory, and he and his crew boarded another helicopter in the flight. This action reduced the number of helicopters from eight to seven.

The next problem the helicopter pilots encountered was a *haboob*—a severe dust storm not uncommon in this part of Iran in the spring. Although the *haboob* caught the pilots by surprise, the flight commander radioed the *Nimitz* and recommended that the mission continue. General Vaught concurred, but these communications could not be heard by the pilots in the other helicopters, because, to minimize the possibility of interception, only the flight commander had a special radio that permitted contact with the *Nimitz*.[48]

Navigating by sight and operating in radio silence, the pilots struggled in the *haboob* conditions. Their options were limited because if they gained more altitude to get out of the storm, they risked being intercepted by Iranian radars. Deciding to remain in the *haboob*, one of the pilots became very disoriented because his gyroscope was not working, and his other navigational devices, while still operable, were being adversely affected by the heat. The pilot could not see any of the other helicopters, and his copilot was feeling sick. Not knowing that one helicopter had already aborted and therefore assuming the mission would not be compromised, the pilot decided to turn back. He also did not know that in twenty-five minutes he would have emerged from the *haboob*. When interviewed afterwards, he stated that if he had known that he was that close to getting out of the *haboob*, and that visibility was normal at Desert One, he would probably have continued his flight. Instead, the number of helicopters was now down to six, the minimum required by the plan.[49]

The six helicopters ultimately arrived at Desert One and were preparing to load the Delta Force operators when one helicopter discovered a fault in its secondary hydraulic system. There was some debate about whether the helicopter could still fly in spite of this defect, but as flight leader, Lieutenant Colonel Seiffert thought that it was too dangerous and that the operation should be aborted. Every ounce of warrior emotion in the hard-charging Beckwith wanted to press on, but as soon as he knew he was down to five helicopters, he told Colonel James Kyle, the Air Force officer designated as commander of the Desert One site, "I can't go forward with five. We gotta go back."[50] Kyle relayed the situation to Washington.[51]

Colonel Beckwith wargamed the new situation in his mind:

> With five helicopters, Delta, minus twenty men, lands at the hide-site in daylight and then the helos fly to their location in the mountains, but hell, we all knew the eccentricities of choppers. There was a good chance two of them would not crank tomorrow. That would leave three helos to pick up 53 hostages, Delta, the [Department of Defense] agents, and the assault team and their three hostages freed from the Foreign Ministry Building. What if one of them got hit with small arms fire as it comes in? That would leave two. Two for 178 people. It was just too close.

As much as Beckwith tried to figure out a way to continue, he could not overcome what he already knew to be the reality. He was not about to ignore all the rehearsals and planning and now "be party to a half-assed loading of a bunch of aircraft and going up and murdering a bunch of fine soldiers."[52] "This is ludicrous," he finally admitted. "It doesn't make sense. Stay with the plan."[53] Colonel Kyle was thinking the same way, telling himself, "I've got to stick with the game plan — we can't go on."[54]

Major General Vaught relayed the recommendation to abort to the White House and to General David Jones, the chairman of the Joint Chiefs of Staff (JCS). Zbigniew Brzezinski, President Carter's national security advisor, weighed the options and considered, "Should I press the President to go ahead with only five helicopters? Here I was alone with the President. Perhaps I could convince him to abandon military prudence, to go in a daring single stroke for the big prize, to take the historic chance."[55] Ultimately, Brzezinski decided to recommend to Carter to continue the operation with five helicopters only if Beckwith agreed.

Beckwith was infuriated by the continued discussion of what he had determined to be inevitable. "I flashed back to the meeting, the one of January 4th, when Pittman and I had recommended we go on with no fewer than six. General Vaught had accepted that recommendation. No more questions remained. It was final!"[56] Indeed, Beckwith acted exactly as the organizational process model would predict and followed established procedure. One scholar of Operation Eagle Claw concludes, "Investigation reveals that Beckwith's decision was strictly in accordance with helicopter abort criteria established during planning and approved by the JCS, JTF, and [National Security Council]. He required six helicopters to continue from Desert One."[57] Likewise, a review group chaired by Admiral James Holloway concluded, "Because only five helicopters were available to proceed against a firm minimum requirement of six, the rescue mission was aborted."[58] In the unfolding crisis at Desert One, Colonel Beckwith was merely implementing a decision that had already been made during the planning process.

Thus, after General Jones confirmed that Beckwith thought the operation was not feasible with only five helicopters, President Carter gave the order to call off the operation and withdraw the force from Iran. The ill-fated mission suffered further catastrophe when, in the course of the evacuation of Desert One, one of the helicopters crashed into a C-130. In the ensuing explosion, and eight military personnel were killed.[59]

Russell Bova notes that the standard operating procedures upon which the organizational process model relies "can be a source of comfort in responding to complex situations, and they can be efficient insofar as they avoid the need to reinvent the wheel every time a new situation develops."[60] Mark Bowden explains Beckwith's abort decision in such a context:

> That was the conclusion the mission planners had reached in advance, after calm, careful deliberation. These automatic-abort scenarios had been predetermined precisely to avoid life-and-death decisions at the last minute. This was clearly an abort situation. On the mission schedule, just after the line "less than six helos," was the word "ABORT," and it was the only word on the page in capital letters.[61]

But Bova continues, noting that standard operating procedures "can also serve as a source of rigidity in dealing with unique circumstances."[62] This potential is alluded to by an intelligence officer associated with Operation Eagle Claw who mused,

> Although it is easy to say in hindsight, the bottom line is that a daring commander in wartime could have and would have continued with five or even four helicopters. Beckwith was a fine Special Forces soldier, but his country was not at war, and his airlift had demonstrated a tendency to break before the first shot was fired. In the middle of the desert, far behind his envisioned time line, and doubtless already concerned about his transportation going into a hide site laager that had never been walked by friendly feet or seen up close by friendly eyes, he sought reassurance from a tired helicopter pilot and a frustrated airfield manager. And he didn't get it. Nor did Major General Vaught order the mission to go forward; and neither did the Chairman of the Joint Chiefs, the Secretary of Defense or the President.[63]

The organizational process model explains the abort decision at Desert One based on either of Bova's possibilities. If Bowden is correct, the decision was the result of prior detailed planning that brought clarity to a crisis and saved lives. If the intelligence officer is correct, the decision was more the result of inflexible deference to an inappropriate frame of reference that stymied initiative in the midst of a new situation. Either way, the decision to abort Operation Eagle Claw was the organizational process model in action.

Example 3: The Federalization of the California Army National Guard During the Los Angeles Riots, 1992

On April 29, 1992, a jury acquitted four Los Angeles police officers accused of beating Rodney King, a black motorist who, while driving under the influence of alcohol, had led California Highway Patrol and Los Angeles Police Department (LAPD) officers on a high speed chase. After he was finally stopped, King tried to assault the police officers, but of more interest to the public was a video taken by a bystander showing several white officers beating a prone and seemingly subdued King with their batons. The acquittal by the jury of ten whites, one Hispanic, and one Filipino sent shock waves through the black community.

The verdicts were announced at 3:15 P.M., and about forty-five minutes later an unruly crowd had gathered at the intersection of Florence and Normandie Avenues in south central Los Angeles, about four miles from the site of the hugely destructive 1965 riot in Watts. By the time police arrived on the scene shortly after 4:30, the mob was assaulting pedestrians, bombarding vehicles with bricks and rocks, and smashing store windows. Rather than responding with a massive show of force, the police retreated from the scene, and the riot continued to grow.

Soon the situation was well beyond anything the police could handle, and at about 9:00 P.M., Governor Pete Wilson ordered the mobilization of 2,000 troops from the California Army National Guard (CANG). Within a few hours of the mobilization order, the CANG established an operations cell at the Los Angeles County Sheriff's Emergency Operations Center (EOC). The sheriff, police, military, and the State Office of Emergency Services were all represented at the EOC, and from that location the CANG received and filled law enforcement support requests.

At this stage of the operation, there was little concern for a specific description of the law enforcement task. The CANG operations cell and the CANG 40th Infantry Division (Mechanized) headquarters merely ordered a subordinate or attached unit to report to a specific law enforcement agency at a particular location and provide whatever support was requested. Los Angeles Police Department district commander Bayan Lewis explained, "The first night the Guard deployed, I would turn to the battalion commander and say, 'I need you to take troops to this location, we need to seal this, we need a barricade on this road, we need so and so.'"[64]

Under this system, the CANG filled requests quickly and effectively. In fact, the CANG was dispatching soldiers to support law enforcement agencies faster than the police could absorb them. Nonetheless, a general mood of panic prevailed, and Mayor Tom Bradley demanded federal troops. Apparently bowing to political pressure, negative news images, and the perception that the CANG had been slow to arrive on the scene, the governor acted on Bradley's request.[65]

At 5:15 A.M. on May 1, President George Bush ordered the deployment of 4,000 federal troops to Los Angeles. One thousand five hundred of these were Marines from Camp Pendleton and the rest were soldiers from the 7th Infantry Division at Fort Ord. As the active duty soldiers were arriving in the area, President Bush announced at 6:00 A.M. his decision to federalize the CANG and place all troops under a central command. This new organization was Joint Task Force–Los Angeles (JTF-LA) commanded by 7th Infantry Division commander Major General Marvin Covault.

Organizational process theory helps explain the decision to federalize the National Guard based on the military's procedural adherence to unity of command, and command and control. FM 100–5, the Army's capstone manual on operations, explained unity of command as "for every objective, ensure unity of effort under one responsible commander."[66] The manual declared it "axiomatic that the employment of military forces in a manner that develops their full combat power requires unity of command." Although it allowed that "coordination may be achieved by cooperation; it is, however, best achieved by vesting a single commander with the requisite authority to direct and to coordinate all forces employed in pursuit of a common goal."[67] Indeed, Major General James Delk, California's deputy adjutant general and commander of CANG forces until federalization, was not surprised by the decision to subordinate the 40th Infantry to JTF-LA. Pointing to a diagram of what became the JTF-LA organization, he said, "This is the standard, out of the manual."[68]

However, this rather fatalistic submission to established organizational routine proved costly in Los Angeles because the federalization of the CANG had an immediate impact on its ability to accept taskings from the local law enforcement authorities. Prior to federalization, junior officers and non-commissioned officers had been permitted great latitude in accepting taskings, and the CANG had been able to accept nearly one hundred percent of the requests it received. After federalization, the CANG began to conduct itself under the rules of the *Posse Comitatus* Act (PCA).

Posse comitatus is Latin for "the power or the force of the county." It refers to the English common law doctrine that empowered the local sheriff to summon men above fifteen years of age to help enforce the law and maintain order in an emergency situation.[69] The 1878 *Posse Comitatus* Act was written in direct response to the use of federal troops to implement the Reconstruction Act of 1867 in the ex-Confederate States, specifically during the hotly-contested 1876 presidential election. During that crisis, President Ulysses Grant sent troops as a *posse comitatus* for federal marshals to use at the polls, if necessary, in South Carolina, Louisiana, and Florida. More broadly, the act reflects the traditional American value of civilian control of the military and a fear of the power of a large standing army.[70]

Posse comitatus prohibits the use of the military to execute the civil laws of the United States, stating, "Whoever, except in cases and under circumstances expressly authorized by the Constitution or Act of Congress, willfully uses any part of the Army or the Air Force as a *posse comitatus* or otherwise to execute the laws shall be fined under this title or imprisoned not more than two years, or both." Significantly, the *Posse Comitatus* Act does not apply to a member of the Reserve Component when not on active federal duty, nor to members of the State National Guard when not in federal service.

After the CANG was federalized, however, all taskings had to be approved centrally in a procedure that took between six and eight hours and involved a review by the JTF-LA commander, the operations officer, and the staff judge advocate. This not only slowed down the process, it also resulted in the CANG being able to accept just twenty percent of all requests. Many observers felt that the full impact of *posse comitatus* was not adequately considered or realized before the decision was made to federalize the CANG.[71]

In reality, because President Bush had cited Chapter 15 of Title 10 of the United States Code in issuing his proclamation to commit federal troops to Los Angeles, he appears to have met the "under circumstances expressly authorized by the Constitution or Act of Congress" of the PCA. Under this interpretation, which is currently widely accepted, PCA restrictions did not apply to the federal troops committed to Los Angeles.[72]

At the time however, the presidential proclamation caused some confusion within JTF-LA. The situation was exacerbated by a certain amount of caution, bewilderment, and overreaction by military attorneys.[73] Reflecting the dominant interpretation that the PCA applied to JTF-LA, General Covault informed Los Angeles Police Chief Daryl Gates and Sheriff Sherman Block that neither the federal soldiers nor the federalized CANG would perform civil law enforcement duties.[74] According to Covault, "It was not the military's mission to solve Los Angeles's crime problem, nor were we trained to do so."[75] One brigade commander in the 7th Infantry Division simply stated, "We weren't going to try to do police work."[76] In his report concerning the military and law enforcement response to the riot, former Director of the Federal Bureau of Investigation William Webster noted, "When missions were received by the US military and requests for the California National Guard were made, the military commanders would consider whether the request was for a 'law enforcement' function or a 'military' function. If the request was for a law enforcement function, the request was uniformly denied."[77] Obviously, this decision reduced the resources available for law enforcement.

A debate would later ensue as to whether or not JTF-LA refused law enforcement missions based on its understanding of the PCA or based on its mission analysis. Major General Delk confessed, "I frankly did not know until several months after the riots that *posse comitatus* did not apply." For his part, Covault claimed, "The JTF-LA commander and his staff understood from the outset the *Posse Comitatus* Act had no effect, and the Act in no way limited the decision-making process within the TF headquarters. The issue in the overall Los Angeles crisis was one of ROLES AND MISSIONS. The military could (and was) best used to create a secure environment, provide widespread presence, and provide a sense of confidence among the populace."[78] Numerous observers, including Director Webster, have called into question Covault's recollection and believe that much of JTF-LA's decision-making was based on a misconception that the PCA limited their actions in Los Angeles situation.[79] Perhaps tellingly, the Army's Center for Army Lessons Learned November 1993 newsletter on the riots is completely silent on the *posse comitatus* issue and the Army's FM 100–19, *Domestic Support Operations*, published in July 1993 after the Los Angeles Riots, is ambiguous. While allowing for exceptions under Title 10 of the US Code, the manual says the military is still only able "to provide limited support to civilian law enforcement agencies (LEAs) indirectly. Under these laws, the military may share certain information and provide equipment, facilities, and other services to LEAs."[80] Such vague wording would hardly be the definitive guidance needed to make decisions such as those required by JTF-LA. Suffice it to say, as Major General Thomas Eres, Staff Judge Advocate for the 40th Infantry Division during the riots, recalls, there was "lots of fog."[81]

While the debate still lingers as to the cause, the ultimate outcome seems clear. In the final analysis, a joint meeting of California law enforcement officers concluded, "The calling in of federal troops appears to have been a mistake. This resulted in the National Guard becoming federalized which severely limited their flexibility and missions they were able to undertake."[82] The decision was an example of the organizational process model's danger of being "a source of rigidity in dealing with unique circumstances."[83] It was also, however, the harbinger of change.

While the Army was still trying to come to grips with *posse comitatus* and other issues from the Los Angeles Riots, Hurricane Andrew struck southern Florida on August 24, 1992. More than 63,000 homes were destroyed and many more were severely damaged. There was virtually no water, electricity, or phone service, and food soon was in short supply. Debris blocked the roads, making movement precarious and unpredictable. The expectation of widespread looting and unchecked criminal activity induced widespread fear.[84]

Some 6,100 members of the Florida National Guard (FLNG) responded quickly, but the scope of the disaster required massive amounts of federal assistance as well. Joint Task Force Andrew was soon formed with Lieutenant General Samuel Ebbesen as commander. By September 11, JTF Andrew consisted of more than 17,000 soldiers, 900 Marines, 3,800 sailors, and 1,000 airmen. The bulk of the soldiers came from the Army's 82nd Airborne and 10th Mountain Divisions.[85]

While the mere presence of the federal troops was viewed as a strong deterrent to crime, JTF Andrew was careful to keep its presence mission within the restrictions of *posse comitatus*. The Army felt that "'patrolling' for the sole purpose of providing security to civilian property would be prohibited."[86] However, patrols that were conducted to perform a humanitarian relief mission such as delivering Meals, Ready to Eat (MREs) were deemed not to violate the PCA even though the mere presence of the patrols deterred crime.[87] In a similar distinction, active soldiers were allowed to set up life support centers to shelter disaster victims who had lost their homes, but were not allowed to provide security at the centers.[88]

Unlike in Los Angeles, the National Guard was not federalized during Hurricane Andrew. Thus, Army Chief of Staff General Gordon Sullivan could state that when federal troops observed criminal activity they could quickly report it and "bring police or members of the National Guard to the scene in a matter of minutes to make arrests or take other law enforcement action."[89]

In addition to providing a presence and reporting criminal activity, military forces attempted to walk a fine line astride other aspects of the thorny *posse comitatus* issue. Federal troops were allowed to direct traffic along roads designated as Military Supply Routes (MSRs), but traffic control on roads that were not MSRs did not pass the "military necessity" test which required the activity be "in furtherance of a military purpose." Thus, Federal soldiers were considered prohibited from enforcing civilian traffic laws under *posse comitatus* and, consequently, most traffic control points were manned by the FLNG.[90] To ensure nuances such as these were accommodated, requests for military support to civilian law enforcement were staffed through the staff judge advocate before JTF Andrew acted on the request.[91]

General Sullivan concludes, "Because of *posse comitatus*, the decision to leave the National Guard under state control was a wise one and should be repeated under similar circumstances in the future."[92] The JTF Andrew After Action Report (AAR) boasts that this decision produced an effective division of labor in which the FLNG was able to conduct the law enforcement mission while the active duty personnel could focus on the relief and recovery mission.[93] The statements of General Sullivan and the AAR reflect organizational learning that occurred after the experience with federalization during the Los Angeles Riots. In spite of their preference for routine, Graham Allison notes that "organizations do change." "Learning and change," he continues, "are influenced by existing organizational capabilities."[94] The military learned in Los Angeles that a National Guard under state control has certain capabilities a federalized National Guard does not. Joint Task Force-Los Angeles accomplished its mission in spite of the federalization decision, but it did so with enough disruption to generate a rethinking of the previously held conventional wisdom of federalizing the National Guard in the name of unity of command.

Utility of the Organizational Process Model

Because governments consist of bureaucracies and organizations, bureaucratic and organizational models will always figure prominently in decisions made about the use of force. The bureaucratic model is most applicable when there is time for political negotiating to take shape.[95] The organizational process model is very appropriate for the myriad other routine decisions that are made by low-level actors who apply general principles in an expeditious way.[96]

This is not to say that the organizational process model is applicable only for short-term decisions. Over time, bureaucracies develop differences not only in their routines but also in their perspectives. The organizational process model also accounts for decisions made by more senior members who apply the organization's general principles and perspectives to long-term issues such as budget making.[97] Likewise, the military has certainly applied lessons learned from Vietnam, including a revised appreciation of counterinsurgency operations that has been codified in doctrine and practice, to its operations in Afghanistan.

Without a doubt, the organizational process model explains many of the decisions that relate to how an organization carries out policies. It is an excellent means of ensuring a standard response by large numbers of ordinary individuals even in complex situations. However, "this regularized capability for adequate performance is purchased at the price of standardization." [98] As Graham Allison notes, "because of standard procedures, organizational behavior in particular instances often appears unduly formalized, sluggish, or inappropriate."[99] In those cases that require more agile, imaginative, and situationally-specific decision-making, the small group model may be more appropriate.

6

The Small Group Model

The small group model offers the policy maker several advantages over bureaucratic and organizational models. According to Robert Wendzel, these include:

- The absence of significant conflict, because there will be few viewpoints to reconcile;
- a free and open interchange of opinion among members, because there will be no organizational interests to protect;
- swift and decisive action;
- possible innovation and experimentation; and
- the possibility of maintaining secrecy.[1]

There are a number of different organizational options for the small group. One is the informal small group that meets regularly but lacks a formal institutional base. Another is the ad hoc group that is created to deal with a specific problem and then disbands once its work is completed. The final type is permanent in nature, has an institutional base, and is designed to perform a set of specified functions.[2]

In all these organizational options, a group of individuals assembles to make a collective choice about alternatives that they generate as a group or that are presented to them. In either scenario, the decision is attributable to the group rather than any of the individuals in it, and therefore this model takes into account social processes in decision-making. One of the main dangers inherent in the social influences associated with small groups is the tendency toward the "groupthink" that results from the strong internal pressure placed on individual members to conform to the evolving group norm.[3] The Blockade Board in the Civil War, the Executive Committee during the Cuban Missile Crisis in 1962, and the Restricted Interagency Group tasked to develop a strategy toward Nicaragua in the 1980s are all examples of small group decision-making.

Example 1: The Blockade Board in the Civil War, 1861

On April 19, 1861, six days after the withdrawal of Federal forces from Fort Sumter, South Carolina, President Abraham Lincoln issued a proclamation declaring the blockade of the Southern States from South Carolina to Texas. On April 27, the blockade was extended to Virginia and North Carolina. The purpose of the blockade was to isolate the Confederacy from European trade. The proclamation read:

> Now therefore I, Abraham Lincoln, President of the United States ... have further deemed it advisable to set on foot a blockade of the ports within the States aforesaid, in pursuance of the laws of the United States and of the Law of Nations in such case provided. For this purpose a competent force will be posted so as to prevent entrance and exit of vessels from the ports aforesaid. If, therefore, with a view to violate such blockade, a vessel shall approach or shall attempt to leave any of the said ports, she will be duly warned by the commander of one of the blockading vessels, who will endorse on her register the fact and date of such warning, and if the same vessel shall again attempt to enter or leave the blockaded port, she will be captured, and sent to the nearest convenient port for such proceedings against her, and her cargo as prize, as may he deemed advisable.

Declaring a blockade and making it effective, however, were two different things. With 189 harbor and river openings along the 3,549 miles of Confederate shoreline between the Potomac and the Rio Grande, clearly some focus was needed. Responsibility for the Federal blockade strategy rested with the Blockade Board (also called the Navy Board, the Strategy Board, and the Committee on Conference) which Secretary of the Navy Gideon Welles created in June 1861 to study the conduct of the blockade and to devise ways of improving its efficiency.

In analyzing the small group model, Barbara Kellerman notes the importance of understanding "the personality and performance of each of [the small group's] members."[4] Indeed, selecting the right members for a small group is critical to its success, and Secretary Welles was fortunate that "personnel of specialized skills were available."[5] He took great care in selecting the Blockade Board's four members, and he choose well. Captain Samuel Du Pont, a professional naval officer and member of the famous Delaware manufacturing family, was its head. Professor Alexander D. Bache, the second member, was superintendent of the Coast Survey and brought to the Board specialized knowledge of the Confederate coast. The third member, Major John Barnard, was an Army engineer who contributed engineering and fortification expertise, and provided some (very) limited liaison between the Army and the Board. Commander Charles Davis was the fourth member and served as the Board's secretary.[6]

Captain Du Pont was an excellent choice as the Board's president, because he was one of the few officers in the Federal Navy who had previous experience with blockading during the Mexican War. Indeed, two days before he arrived in Washington to assume his duties on the Board, Du Pont recalled his previous blockading experience in a letter to a friend: "During the Mexican War I had two hard years' work at it, with endless correspondence with naval and diplomatic functionaries, for I established the first blockade on the western coast." In Mexico Du Pont had learned firsthand the practical difficulties of a blockade: Such an undertaking is very resource intensive in terms of both ships and supplies. To maintain the blockade, ships cannot abandon their positions to get supplies. Either the resupply points must be close enough to facilitate the blockade or additional ships must come in to replace the departing ones. Du Pont would carry these lessons with him to the Blockade Board.

Captain Du Pont also had experience as an effective administrator, having served on the Lighthouse and Efficiency Boards in the 1850s. Yet in spite of his impressive personal credentials, Du Pont ran the Board without stifling creative thought, taking full advantage of the talents of the Board's other members.[7] Kellerman notes the importance of "the identity of the leader and his role in the group,"[8] and Du Pont's biographer Kevin Weddle declares that "the combined effects of [Du Pont's] experiences had produced an officer uniquely qualified for the chairmanship of the Blockade Board."[9]

Professor Bache was the great grandson of Benjamin Franklin and, like his famous ancestor, was one of the scientific luminaries of his day. Bache graduated first in his West Point Class of 1825 and then taught mechanical engineering at his alma mater before serving as an engineer during the construction of Fort Adams at Newport, Rhode Island. He resigned his military commission in 1828 when he was appointed professor of natural philosophy at the University of Pennsylvania and became editor of the *Journal of the Franklin Institute* a year later. He quickly used his position to raise the standards of American science and establish a professional science community. On the strength of this stellar reputation, Bache became the second superintendent of the Coast Survey when Ferdinand Hassler died in 1843.[10]

The Coast Survey was technically a temporary governmental agency under the aegis of the Treasury Department. It had the rather modest mission of gathering information in the forms of hydrography, geodesy, topography, and the printing arts to aid the navigator. Bache interpreted his mission much more broadly and expanded the Coast Survey's core mission. He began studying the Gulf Stream Current in 1845, directed tide observers to make meteorological observations, began geomagnetic studies, and guided the development of mathematical techniques for a variety of scientific applications. He used his vast network of old Army friends, such as the distinguished engineer Brigadier General Joseph Totten, to augment his own scientists and surveyors, and made himself valuable as a member of several ad hoc bodies and boards such as the Lighthouse Board.[11]

In spite of this success, Professor Bache "was a frightened man in early 1861." The sectional crisis not only threatened the country; the loss of access to the thousands of miles of coastline in the seceded states threatened the Coast Survey as well. In January 1861, he confided to a friend, "the terrible disruption of our country ... will sweep our organization away entirely, or sadly cripple it." Indeed, it was this concern for the Coast Survey's survival that led Bache to first propose the idea of a Blockade Board. As Kevin Weddle notes, "In this respect, Bache was no different from any other government bureaucrat; he was determined to protect his agency from any threat by proving it was indispensable." This is not to say Bache's motivation was purely based on self-interest. He was also a dedicated patriot who wanted to contribute to the war effort.[12] Nonetheless, he illustrates Kellerman's observation that the reality of "where you stand depends on where you sit" that Graham Allison applied to the bureaucratic politics model is present in the membership of small groups as well.[13]

Bache's membership on the Blockade Board was a win-win situation. It certainly secured the future of his beloved Coast Survey, but his technical contribution and expertise was also critical to the Board's success. Without Coast Survey maps, an effective blockade of the Southern ports would have clearly been impossible.[14] Weddle declares that Bache's "contributions proved indispensable."[15]

Bache had originally envisioned his friend Totten as a fellow Board member, but because Totten's duties prevented him from participating, Secretary of War Simon Cameron nominated Major John Barnard instead. Barnard was the engineer in charge of the defenses of Washington and an expert on coastal topography. After graduating from West Point in 1833, he spent the next twenty-eight years in the Corps of Engineers and had been the superintendent of the Military Academy. He was an expert on the construction of coastal defenses and harbor improvements which, along with his writings on scientific and engineering subjects, brought him in contact with Bache. Indeed, the pair had worked together on the construction of the defenses around Washington. Barnard brought his valuable expertise on coastal defenses to the Board, but his presence as an Army officer should not be misconstrued as giving the Board the characteristics of a modern "joint staff."[16] Indeed, whatever liaison he provided between the Board and the Army was largely tangential.

Rounding out the Board was Commander Charles Henry Davis, another of Bache's close friends. Davis had joined the Navy as a midshipman in 1817 and had served with Bache on detached duty with the Coast Survey in the 1850s and on various boards. Davis was now a staff officer with the Bureau of Detail in Washington, DC. What he lacked in sea-going experience, Davis made up for as a scientific expert and with his engaging personality. His credentials included having been the head of the Naval Almanac, an agency located in Cambridge, Massachusetts, that was related to, but separate from, the Naval Observatory. The Almanac produced navigational and astronomical tables that would be valuable to blockade planning. Moreover, Davis was a life-long friend of Du Pont, who thought so highly of Davis that he believed (correctly) he would one day head the Naval Observatory.[17]

Kellerman notes the importance of "the history and nature of the small group." The degree to which members have formed assessments of and attitudes toward one another, share goals and interests, and have a tradition of working together greatly impacts the decision-making process in the small group model.[18] Certainly the members of the Blockade Board had the group dynamic necessary to be successful, and their synergy would sustain them in trying times. The Board kept a demanding schedule "in the pressure cooker atmosphere of wartime Washington" where all its members save Captain Du Pont also had non-Board duties as well. Nonetheless, the men "quickly bonded into a tight-knit group" and "forged close personal bonds." They routinely "dined and socialized together."[19]

Secretary Welles's initial guidance to the Board was expansive. He instructed Du Pont,

> The Navy Department is desirous to condense all the information in the archives of the Government which may be considered useful to the Blockading Squadrons; and the Board are therefore requested to prepare such matters as in their judgment may seem necessary: first, extending from the Chesapeake to Key West; second, from Key West to the extreme Southern point of Texas. It is imperative that two or more points should be taken possession of on the Atlantic Coast, and Fernandina and Port Royal are spoken of. Perhaps others will occur to the board. All facts bearing on such a contemplated movement are desired at an early moment. Subsequently, similar points in the Gulf of Mexico will be considered. It is also very desirable that the practicability of closing all the Southern ports by mechanical means should be fully discussed and reported upon.[20]

Welles was clear that he expected the Board to tackle two of the blockade's key challenges: a lack of local information and a lack of logistical bases. Welles had also ordered the Board to plan for the seizure of additional bases, first in the Atlantic and then in the Gulf of Mexico. It was a far-reaching task, but Welles's guidance was clear and helpful.[21]

The Board held its first meeting on June 27, 1861. Du Pont's goal was to determine how the blockading squadrons should best execute their missions. He was shocked that squadron commanders seemed content to merely cruise aimlessly up and down the coast with a few vessels. Du Pont knew a system was required; what Bache called a "manual" for blockading. Du Pont also recognized the need for joint expeditions to capture logistical bases both to support the blockade and to be used as springboards to launch ground operations into the Confederate interior. The Board discussed the need for ground troops to seize and hold such bases. Du Pont had clearly defined the Board's work as being to provide the necessary operational and strategic direction for the blockade and its supporting joint operations.[22]

The Board presented its first two reports to Secretary Welles on July 5 and 13. The first report confirmed the need for additional bases, stating, "It seems to be indispensable that there should exist a convenient coal depot on the southern extremity of the line of Atlantic blockades ... [and it] might be used not only as a coal depot for coal, but as a depot for provisions and common stores, as a harbor of refuge, and as a general rendezvous, or headquarters, for that part of the coast."[23] Fernandina, Florida was the Board's recommendation to meet this requirement.

The second report focused on the need for a second base farther north. First, the Board recommended closing the inlets between the Cape Hatteras barrier islands. Then it examined three potential bases along the South Carolina coast: Port Royal Sound, Bull's Bay, and Saint Helena Sound. Seizing a base deep in the South would be risky and would require a formidable ground and naval force, but the strategic payoff would be great. Although the Board recognized the superiority of the harbor at Port Royal Sound, it also assumed the Confederates would mount a difficult defense there. Thus, the Board recommended seizing Bull's Bay.[24]

The Board issued two more reports on July 19 and 29. Perhaps the most important of the many recommendations in these reports was that responsibility for the Atlantic blockade be divided between two squadrons. This arrangement would streamline command and control, and reduce the burdens placed on the commanders. Later, the Board would recommend the Gulf Blockading Squadron also be divided into two separate commands.[25]

On August 6, the Board issued its first report on the Gulf of Mexico. The geographic complexities of the Mississippi River Delta made this region particularly difficult to blockade, and the Board was quick to point out that "the blockade of the river ... does not close the port [of New Orleans]." Because the capture of New Orleans would require such a large naval and military force, the Board recommended action against New Orleans be delayed until "we are prepared to ascend the river with vessels of war sufficiently protected to contend with the forts." In the meantime, the Board recommended seizing Ship Island, a barrier island midway between New Orleans and Mobile. Ship Island would serve as the headquarters and logistical base for the Gulf Blockading Squadron and would be useful as a jumping off point for any future attack against either New Orleans or Mobile.[26]

Captain Du Pont's stellar work on the Board catapulted him ahead of several more senior officers when it came time to select a commander for the important Port Royal Expedition. It also caused him to divide his attention between the Board and his sea command, and it was not until September 3 that the Board completed its second Gulf report. This report summarized the geography and topography of the rest of the Gulf, including the Florida Keys and the entire coast of Texas. Finally, on September 19, the Board made its last report which supplemented the first Gulf report by carefully outlining the defenses of Ship Island. With his new appointment as Flag Officer, Du Pont was now fully engaged in his Port Royal Expedition duties, but he asked the Department of the Navy to allow the Board to make one more report—the manual for the conduct of blockading. Secretary Welles failed to act on Du Pont's request, and this report was never finished.[27]

The Blockade Board was one of several instances in which Secretary of the Navy Gideon Welles turned to the small group model to help generate options (Library of Congress).

The Blockade Board was a resounding success, producing what Weddle describes as "a military (naval) strategy that was fully coordinated with the national strategy and government policies."[28] Indeed, the Department of the Navy accepted most of the Board's recommendations. Welles split the Atlantic Blockading Squadron into the North and South Atlantic Blockading Squadrons, commanded respectively by Flag Officer Louis Goldsborough and Du Pont. Likewise, the Gulf Blockading Squadron was divided into the East and West Gulf Blockading Squadrons under Flag Officers William McKean and David Farragut, respectively. The Lincoln Administration and the War and Navy Departments also took swift action on the Board's recommendations for joint operations, seizing Hatteras Inlet in August 1861, Port Royal and Ship Island in November, and Fernandina in March 1862. Secretary Welles also used the model of the Blockade Board to establish other boards and commissions such as the Board of Ironclad Vessels and the Board of Naval Examiners. Finally, the Board succeeded in its mission of condensing the wealth of information on the Confederate coast into a useable form that was readily available to the squadron commanders. In praising the work of the Blockade Board, Weddle argues that "the Civil War saw no comparable organization, staff, or agency that systematically formulated naval or military strategy."[29] Du Pont and his colleagues had done their work well.

The initial planning effort by the Blockade Board was deliberate and detailed, but as the campaigns progressed, planning became much more haphazard. In September 1862, Du Pont, now a rear admiral, pleaded with Assistant Secretary of the Navy Fox to not "go it half cocked about Charleston—it is a bigger job than Port Royal.... You & I planned the first ... let us consult together again."[30] Instead of the careful planning of the Blockade Board, Du Pont lamented that now the "desire of the President and others *to strike a blow* somewhere" was not accompanied by having "someone [who] would sit down and study how the blow was to be given." The result, according to Du Pont, was that the April 7, 1863, attack on Charleston was a "chaotic conception" rather than the result of a military plan.[31] For Du Pont, the attack's defeat was predictable.

Admiral Du Pont's criticism of the decline in the quality of the Federal planning effort shows the criticality of determining the organizational type of the small group. The situation required a group that would remain active for the duration of the war. Instead, by October 1861 the Blockade Board ceased to function, long before its work was complete and unfortunately "leaving its early promise ... never fully realized."[32] Donald Stoker correctly asserts that the Board's "continuance would have been a good thing for the navy."[33] In spite of its premature demise, however, the Blockade Board showed the advantages a small group can offer when innovation is required.

The Blockade Board also illustrates the utility of the small group model as a more flexible option to bureaucratic and organizational process decision-making. Weddle calls it "one of the most interesting historical ironies of the war that the Union army, with a well-developed bureaucracy, a body of strategic writing and theory, and a general-in-chief, was unable to formulate a coherent military strategy until the war was almost three years old. On the other hand, the US Navy, with none of the army's advantages, developed a superb strategic concept in less than three months that lasted, with few changes, until the end of the war."[34] The Blockade Board is an excellent example how a small group with the right membership, leadership, and structure can yield swift and effective decisions.

Example 2: The Executive Committee During the Cuban Missile Crisis, 1962

On October 22, 1962, President John Kennedy issued National Security Action Memorandum 196. It stated:

> I hereby establish, for the purpose of effective conduct of the operations of the Executive Branch in the current crisis, an Executive Committee of the National Security Council. This committee will meet, until further notice, daily at 10:00 A.M. in the Cabinet Room. I shall act as Chairman of this committee, and its additional regular members will be as follows: the Vice President, the Secretary of State, the Secretary of Defense, the Secretary of the Treasury, the Attorney General, the Director of Central Intelligence, the Under Secretary of State, the Deputy Secretary of Defense, the Chairman of the Joint Chiefs of Staff, the Ambassador-at-Large, the Special Counsel, and the Special Assistant to the President for National Security Affairs.[35]

The "current crisis" to which the memo refers is the October 16 discovery of Soviet missiles in Cuba. President Kennedy turned to this newly created "Excomm" to generate options and make recommendations for action. Indeed, the small group model, with its capacity for secrecy, speed, and rapid response, is often used in such crisis situations. Moreover, by the specific way in which Excomm was to be conducted, Kennedy sought to correct the dismal decision-making process that had led to the Bay of Pigs fiasco the previous year.[36]

Research psychologist Irving Janis believes President Kennedy accomplished this goal with Excomm, declaring it "did everything right from thoroughly canvassing a wide range of alternative courses to making detailed provisions for executing the chosen course of action."[37] Janis is particularly emphatic in his argument that Excomm avoided the groupthink phenomenon so dangerously common to small group decision-making. He describes groupthink as "a mode of thinking that people engage in when they are deeply involved in a cohesive in-group, when the members' strivings for unanimity override their motivation to realistically appraise alternative courses of action."[38] It is "the tendency for groups to reach decisions without accurately assessing their consequences, because individual members tend to go along with ideas they think the others support."[39] This tendency is generated by the individual's preference to take the easy route and avoid being criticized and alienated for advocating a position that is contrary to that of other members. The result is "close-mindedness or a collective reluctance to question basic assumptions about the problem at hand," resulting in a "shared illusion" of consensus within the group.[40]

According to Janis, the seven major pitfalls that generally afflict group problem solving are:

- discussions are limited to a few alternative courses of action without a survey of the full range of alternatives;
- the group does not survey the objectives to be fulfilled and the values implicated by the choice;
- the group fails to reexamine the course of action initially preferred by the majority of the members from the standpoint of nonobvious risks and drawbacks that had not been considered when it was originally evaluated;
- the members make little or no attempt to obtain information from experts who can supply sound estimates of losses and gains to be expected from alternate courses of action;
- selective bias is shown in the way the group reacts to factual information and relevant judgments from experts, the mass media, and outside critics; and
- the members spend little time deliberating about how the chosen policy might be hindered by bureaucratic inertia, sabotaged by political opponents, or temporarily derailed by the common accidents that happen to the best of well-laid plans.[41]

Janis claims Excomm successfully avoided these dangers thanks to four specific procedural innovations:

- new definitions of the participants' roles;
- changes in group atmosphere;
- meetings of subgroups; and
- leaderless sessions.

As a result of these changes, Janis contends "the members of the Executive Committee avoided succumbing to groupthink, despite the fact that they formed a cohesive group with all the usual social pressures operating to induce conformity with group norms."[42]

The Executive Committee's members ranged from those who participated because of their position (Secretary of State Dean Rusk and Secretary of the Treasury Douglas Dillon), to those who had a particular expertise (Assistant Secretary of Defense Paul Nitze and Soviet expert Llewellyn Thompson), to those who had a special relationship with the president (brother Attorney General Robert Kennedy and White House staff member Theodore Sorensen). Indeed, Sorenson noted the members "had little in common except the president's desire for their judgment."[43] Still, they knew each other fairly well, and five had been part of the Bay of Pigs decision-making process.[44]

The Executive Committee used a series of small group meetings like this one on October 29, 1962, to help resolve the Cuban Missile Crisis (photograph by Cecil Stoughton; John F. Kennedy Presidential Library and Museum).

Unlike in previous situations, however, each member of Excomm was expected to function as a skeptical "generalist." Indeed, after the Bay of Pigs, President Kennedy had become wary of military and CIA "experts." In this new role, Excomm members were supposed to participate not primarily as representatives for their respective agencies but as critical thinkers. Thus, they were encouraged to make comments outside their presumed area of expertise. Of particular note were the roles entrusted to Robert Kennedy and Sorensen. These "intellectual watchdogs" were told to ruthlessly explore every point of contention to ensure thorough analysis. Kennedy in particular acted as devil's advocate to argue on behalf of a contrary position.[45]

Controlling a group's formal decision rules and its agenda provides a common means of structuring the terms of debate in a way that can lead to groupthink.[46] Instead, the Excomm group atmosphere emphasized frank and free-wheeling discussion, without a formal agenda. President Kennedy did set early limits on the scope of Excomm's deliberations, rejecting diplomacy as an option and insisting that forceful action must be taken. Within those limits, however, it was up to the group to solve the problem, and Kennedy told Excomm to make an "intensive survey of the dangers and all possible courses of action." New advisors, men in secondary positions, and subject matter experts were frequently brought into the deliberations to obtain fresh views and new information. All speakers were questioned closely, and, to help overcome any hesitancy to speak, visitors were explicitly asked for their opinions.

Formal protocol was also suspended. Although President Kennedy nominally chaired the meetings, he "led with a light hand."[47] Once the intelligence reports that opened each meeting were finished, discussion became fully egalitarian and informal. For instance, Excomm members were not required to ask Kennedy's permission to speak (although they sometimes did). They regularly challenged one another, including Kennedy, and explicit shows of deference were infrequent.[48] As Deputy Secretary of Defense Roswell Gilpatric recalled, Kennedy "not only allowed, he encouraged open discussion. There was never any attempt to cut off interventions."[49] Similarly, Richard Bissell, the CIA's Deputy Director of Plans, felt that "under Kennedy, there was more opportunity — and indeed, more incentive — for the individuals in the EXCOM[M] to say what they really thought, instead of trying to agree on a watered-down result."[50]

Group members were also allowed to take a position, but then retreat from it and change their mind if they later decided their initial choice was wrong. Rather than a sign of waffling or weakness, this group dynamic was an important means of encouraging debate. As examples, McNamara changed his mind from wanting to do nothing to supporting a blockade, and Dillon was turned against the air strike option by Robert Kennedy's characterization of a surprise attack as a Pearl Harbor-like betrayal of American moral tradition. In fact, it was only after hearing the arguments of McNamara and others that President Kennedy changed his own mind from surgical air strike to blockade.[51] These examples are not unique. Robert Kennedy recalled that no member of Excomm "was consistent in his opinion from the very beginning to the very end."[52]

To facilitate critical thinking, Excomm sometimes broke into two groups that would meet independently to discuss a subject. Then Excomm would reassemble to debate and cross-examine each subgroup's conclusions. The members of the White House staff also broke off and met separately with the president, "away from the inhibiting pressure of the grandees in the Cabinet Room."[53]

President Kennedy would sometimes absent himself from Excomm meetings to avoid unduly influencing the discussion. He was particularly careful to do this during the preliminary phases when the full range of options was being discussed for the first time. Robert Kennedy saw the wisdom of his brother's absences, noting that "personalities change when the President is present, and frequently even strong men make recommendations on the basis of what they believe the President wishes to hear."[54] He felt "there was less give and take with the President in the room," and that "there was the danger that by indicating his own view and leanings, he would cause others just to fall in line."[55] It should be noted, however, as Barbara Kellerman does, that "although democratic in leadership style, and able to delegate responsibility for gathering and evaluating information, [Kennedy] was clearly in command."[56]

Of course, the evolution of scholarship being what it is, subsequent observers have challenged Janis's conclusion that Excomm avoided the dangers of groupthink. David Gibson rather modestly distinguishes that "we might say instead of groupthink, the choice rested on group*talk* — on most ExComm members converging on the same story while preventing its discontents from raising an effective objection."[57] Robert Thompson is more critical, saying that Kennedy purposely assembled a "controllable group that represented political support for the action he craved."[58] More directly, Glenn Hastedt reports that Excomm's deliberations reflected at least three decision-making defects that are fully consistent with groupthink. These, he argues, are its narrow mandate to consider only coercive measures, ostracizing members who sought to expand the list of options under consideration and break out of the group consensus, and Sorenson and Robert Kennedy's acting as surrogate leaders for President Kennedy to limit the choice of policy alternatives and to stifle discussion.[59]

These arguments notwithstanding, the results seem to speak for themselves. Nuclear holocaust was avoided, and the Cuban Missile Crisis was resolved according to terms favorable to the United States. As Janis deadpans, "the United States government must have done something right."[60] On the other hand, Janis worries "that if groupthink tendencies had become dominant, [Excomm] would have chosen a more militaristic course of action and would have put it into operation in a much more provocative way, perhaps plunging the two superpowers over the brink."[61] The Executive Committee served President Kennedy well and is a positive example of small group decision-making.

Example 3: The Restricted Interagency Group in Developing a Strategy Toward Nicaragua, 1980s

President Ronald Reagan's resolve to undermine the Soviet system exceeded that of any previous administration.[62] A specific global objective outlined in his National Security Decision Directive 32 was "to contain and reverse the expansion of Soviet control and military presence throughout the world, and to increase the costs of Soviet support and use of proxy, terrorist and subversive forces."[63] Reagan's strategy to aid anti–Soviet insurgencies attempting to overthrow Marxist regimes in the Third World eventually became known as the Reagan Doctrine.[64] Specifically, he saw Nicaragua's increasing ties with the Soviet Union, East Germany, and Cuba, as well as Nicaragua's growing military, as a serious threat to American interests in Central America.[65] Central Intelligence Agency Director William Casey shared Reagan's concern, and in a March 1980 meeting of the National Security Council, Casey asked, "If we can't stop Soviet expansion in a place like Nicaragua, where the hell can we?"[66]

President Reagan soon embarked on a campaign to change the US approach to Nicaragua from one of the moderation that characterized his predecessor Jimmy Carter's policy to one of confrontation. Reagan pulled together representatives from such organizations as the CIA, the Department of Defense, the National Security Council (NSC), and the Department of State to create a secret planning body called the Restricted Interagency Group (RIG) to develop options. These meetings were chaired by the State Department representative who was first Thomas Ender, then Langhorne Motley, and finally Elliott Abrams. Other members included General Paul Gorman, the commander of US Southern Command; Deputy Secretary of State for Inter-American Affairs Nestor Sanchez; and CIA operations officer Duane Clarridge.[67]

Within the RIG there was a three-man core group that was particularly important. It included Abrams, Lieutenant Colonel Oliver North of the NSC, and Alan Fiers, the chief of the CIA's Central America Task Force. As a testimony to this core group's importance and activity, Abram's log shows that it met seven times during the years 1985 and 1986. During that same period, the full RIG met eighteen times.[68]

Much of the work of the RIG was forthright and subjected to legitimate administration review. For example, on February 7, 1984, President Reagan signed National Security Decision Directive 124 titled "Central America: Promoting Democracy, Economic

Improvement, and Peace." The directive approved "measures from the Action Plan outlined in the Restricted Interagency Group Report 'Where Next in Central America.'"[69] Among the taskings was that "The Secretary of Defense, in coordination with the Secretary of State and the Director of Central Intelligence, should conduct US military activities in the region that are sufficient to reassure our friends and enhance our diplomatic efforts." There was specific direction to "develop and implement plans for new exercises in Honduras and naval activities in waters off Central America in a manner that will maintain steady pressure on the Nicaraguans and deter Nicaraguan military actions against its neighbors."[70]

Military exercises already comprised an important part of the US effort to influence Nicaragua without resorting to force. When Nicaragua's Sandinista President Daniel Ortega rebuffed the entreaties of Assistant Secretary Enders during negotiations at Managua in August 1981, the US responded with Exercise Halcon Vista (Falcon View) in October. This relatively low-key affair was a three-day exercise designed for the immediate military purpose of evaluating the US-Honduran ability to "detect and intercept hostile coastal incursions."[71] Strategically, the exercise was "designed to signal that the United States was in a position to intervene militarily if the FSLN [the Nicaraguan communist party] did not acquiesce" to Ender's proposals.[72]

Halcon Vista was followed by Combined Movement, which began in late July 1982 and lasted two weeks. The object of these maneuvers was to "conduct combined/joint movement in support of Honduran Army forces to meet an aggressor force in a remote area of Honduras."[73] Then came Big Pine in February 1983 and the expansive Big Pine II, which began on August 3, 1983 and lasted six months. Big Pine II involved 12,000 US troops and included drills in naval interdiction, aerial bombings, airlifts, amphibious landings, and counterinsurgency techniques.[74] A major part of these exercises were engineering projects that built or improved air strips and other infrastructure. As a July 23 article in the New York *Times* explained, "The plan approved by Mr. Reagan does not envisage any immediate combat role for United States forces, but does call for making preparations so that American forces can be swiftly called into action if necessary."[75]

The size and pace of the exercises grew after National Security Decision Directive 124. These later exercises included the three-month Grenadero I in April 1984; Big Pine III, involving 4,500 troops from January to April 1985; the 6,000-man Universal Trek '85 from April to May 1985; and Solid Shield in May 1987, which involved 50,000 US personnel.[76] This almost continuous US presence in the region was an effective and legitimate means of waging psychological war on the Sandinistas and building the infrastructure to make overt military intervention possible and the threat credible.[77]

These exercises proved critical when, after Congress voted to terminate aid to the anti–Sandinista force known as the Contras on February 3, 1988, Nicaragua launched an attack into Honduras the next month. The US responded by deploying a task force comprised of 2,000 troops from the 82nd Airborne Division at Fort Bragg, North Carolina, and 1,100 more from the 7th Infantry Division (Light) at Fort Ord, California. The deployment was codenamed Operation Golden Pheasant, and White House spokesman Marlin Fitzwater described it as "a measured response designed to show our staunch support to the democratic government of Honduras at a time when its territorial integrity is being violated by the Cuban and Soviet-supported Sandinista army."[78]

Among the facilities used for the deployment was the Palmerola air force base where construction as a military command and control facility had begun in 1983, and after Big Pine II a contingent of troops was left behind there to "aid in the operational aspects" of future maneuvers. By March 1984, a month after President Reagan had signed National Security Decision Directive 124, 1,486 US troops manned Palmerola, comprising "a self-contained combat control team fully able to direct a battle force of tens of thousands of troops."[79] Although Operation Golden Pheasant was not nearly that ambitious, the presence of the Palmerola infrastructure greatly facilitated the operation.

In spite of being accompanied be a certain amount of cynicism and controversy, Operation Golden Pheasant was effective. An article in *Military Review* boasted,

> If Exercise Golden Pheasant's demonstration of the rapid deployment of 3,000 soldiers provided a lesson for Americans, it was this: when the exercise began more than 2,000 Sandinistas were inside Honduras and showed no signs of leaving. After the deployment of US troops, the Sandinistas not only withdrew from the country, but for the first time, began to engage in serious negotiations for peace. Golden Pheasant was a show of force that worked.[80]

It was the fruit of National Security Decision Directive 124s directive that "Emergency Deployment Readiness Exercises (EDRE) should be conducted in Honduras commencing in March and later in 1984, as necessary, to demonstrate of our commitment and resolve."[81]

On the other hand, House Speaker Jim Wright accused administration officials of "obviously trying to do everything in their power to keep the war going," and Senator Christopher Dodd, the most persistent Democratic critic of Reagan's Central American policy, suggested the whole operation was designed to trick Congress into approving a new aid package. Following Dodd's theory, *Time* argued that by deploying troops, President Reagan sent a "clear, if unspoken, message to the US public: if Congress refused to fund the contras' fight against the Marxist-oriented Sandinista regime, the American boys just might have to do the job instead." Senator Tom Harkin went one step further, accusing the administration of pursuing a "three-for-one strategy" designed to revive US aid to the Contras, to sabotage the Central American peace talks, and to divert public attention from Independent Counsel Lawrence Walsh's criminal indictments of National Security Advisor John Poindexter and Lieutenant Colonel North and their accomplices in what became known as the Iran-Contra Scandal.[82]

This Iran-Contra Scandal was the unintended result of President Reagan's decision, based on input from the RIG, to challenge the Sandinistas through an insurgency.[83] The RIG had generated an options paper for discussion by the NSC at a meeting on November 16, 1981. The paper included such alternatives as military action against Cuba, support for interdiction forces in Nicaragua, cooperation with Argentina for paramilitary insurgent action, and expansion of political aid to internal anti–Sandinista groups.[84] As a result of the NSC meeting, President Reagan issued National Security Decision Directive 17 on "Cuba and Central America" on January 4, 1982. It included decisions to "provide military training for indigenous units and leaders both in and out of country" and to "support democratic forces in Nicaragua."[85]

Such a course was facilitated by the presence of the small bands of *Guardia Nacional* that had taken refuge in neighboring Honduras and Guatemala following the collapse of the Somoza regime in Nicaragua. Sometime around April 1981, Honduran Police Chief and Army Colonel Gustavo Alvarez had presented CIA Director Casey with a proposal to transform the anti–Sandinista exiles in Honduras into a force potent enough to launch into Nicaragua to ignite a civil war. Alvarez surmised the action would likely prompt a Nicaraguan retaliatory strike into Honduras to which the US could respond with a crushing invasion to solve the Nicaraguan problem once and for all.[86]

The plan seemed perfect to Casey, and he took it to President Reagan, emphasizing that the US would merely be "buying in" to an existing operation.[87] Reagan liked the idea and authorized the expansion of the heretofore small covert insurgent aid program to $19 million. With this support, the Contras, as the insurgent movement became known, eventually grew to a strength of some ten thousand.[88]

While Reagan was completely convinced of the legitimacy of supporting the insurgency, the opinion was far from universal. For example, in March 1986, the House voted against the Reagan Administration's $100 million military and humanitarian aid package for the Contras.[89] Indeed, "the high priority assigned the issue by the president and the intense scrutiny given the policy by Congress" would be a continual source of friction between the two branches of government.[90] The Sandinistas were well aware of the divisiveness of the issue in the US and used the fickle nature of US support to gauge the intensity of their activities. In fact, the inconsistency of congressional support led the Reagan Administration to take extraordinary measures to keep the insurgency alive; measures that pushed the very limits of the operation's domestic legitimacy.

In an effort to circumvent the sporadic congressional funding, Poindexter and North had initiated a program of "overcharging" Iran for weapons and diverting some of the proceeds to the Contras. On November 21, 1986, President Reagan and Attorney General Edwin Meese made the embarrassing announcement of this scheme, and Reagan dismissed Poindexter and North. As Secretary of Defense Caspar Weinberger put it, "The Contras should have been funded, but there is only one way to secure legal spending by our Government, and that is by vote of the Congress."[91] Because this procedure was not followed, for many, the Iran-Contra Scandal became the defining moment of the Reagan Administration's Nicaragua policy, if not Reagan's very presidency.

All the nuances of Iran-Contra are beyond the scope of this discussion. What is germane is that this blatant departure from the institutional bureaucratic procedure and Lieutenant Colonel North's disproportionate influence represents a danger inherent in small group decision-making. One explanation of how a relatively low-level member of the RIG could gain such a dominant role is that before the arrival of Abrams, "constant feuding among RIG members ... eventually led to a situation in which power gravitated to North."[92] Armed with a charismatic personality, enthusiastic commitment to the cause, and a willingness to portray his individual actions as having more authorization by and knowledge of his superiors than they actually did, North quickly asserted himself. Indeed, Barbara Kellerman notes that in a small group "an unofficial leader may emerge because his traits meet the group's special needs in a given situation."[93] Obviously, this leadership can take on both positive and negative attributes. In North's case, he was so confident in his position that at one point he responded to a colleague's concern that Secretary of State George Shultz might object to a RIG plan to increase anti–Sandinista action by saying, "f*** the Secretary of State."[94]

While North emerged as the most aggressive member of the RIG, US Ambassador to Costa Rica Lewis Tambs insisted that he received orders not from North solely, but from the RIG itself—specifically what he called the "triumvirate" of Abrams, North, and

Fiers. Tambs resigned his post in January 1987 amid reports that he and his staff had improperly assisted the Nicaraguan rebels.[95]

Ambassador Tambs said that when he arrived in Costa Rica in July 1985, Abrams and the other officers of the RIG asked him to persuade the Costa Rican government to allow Contra supply pilots to use a secret 1.2 mile-long airstrip that North's operatives had built just south of the Nicaraguan border. Tambs also reported that he and the chief of the CIA station in Costa Rica were directed to give logistical help to the pilots. Tambs claimed he did not know what was on the flights, but a plane shot down in October was found to be carrying munitions.[96]

Ambassador Tambs also reported the RIG directed him to help with the opening of a new Contra offensive on Nicaragua's southern border with Costa Rica, and that, in pursuit of these orders, he and his staff had succeeded in building a "southern front" of between 1,600 and 2,800 men by September. He then claimed RIG members asked him to persuade the rebels on the southern front to join forces with the Contras based in Honduras. Tambs said he resisted that idea because he did not think the main Contra elements would keep the southern forces well supplied. Other than practical objections to such logistical concerns, however, Tambs said he did not question the legality of what he and his staff were being asked to do. "It was up to the RIG to tell us what the margins were," he explained. "It was complicated, and how were we supposed to know? We didn't have a legal staff at the embassy. If orders came from Washington, I assumed they were legal."[97]

This lack of questioning and the assumption that the covert operation in Nicaragua and the associated Iran-Contra connection was the proper course has been labeled "a spectacular case of groupthink."[98] At the highest levels, President Reagan's advisors succumbed to groupthink by failing to advise him against proceeding "on a highly questionable course of action even in the face of his strong conviction to the contrary."[99] Instead, "they formed cohesive groups, shut out dissenters, and generally presented a united front to the president."[100] While groupthink certainly explains part of the Nicaraguan policy decision-making process, equally if not more instructive is an analysis of Kellerman's argument that "the pattern and structure of its decision-making activity" affects the content of the small group's final choice.[101]

The Tower Commission, a three-man review board Reagan appointed to investigate the Iran-Contra affair, made specific notice of the president's delegative personal management style, but also asserted that this preference did not relieve Reagan from responsibility for supervising the process to ensure "the NSC system did not fail him." Likewise, the Commission commented on the responsibility of the president's advisors, with their knowledge of his delegative style, to be "particularly mindful of the need for special attention to the manner in which [the] arms sale initiative developed and proceeded." The Commission found that neither the president nor his advisors fulfilled these responsibilities, with the result being that "the NSC process did not fail, it simply was largely ignored." Because too much power was delegated to Lieutenant Colonel North and other members of the RIG, the Nicaraguan policy unfolded without the benefit of the formal and systematic review, oversight, and periodic evaluation it required. As Kellerman would predict, the overly decentralized and unstructured nature of the small group involved in the process resulted in what the Tower Commission deemed "an unprofessional and, in substantial part, unsatisfactory operation."[102]

Utility of the Small Group Model

Small group decision-making is a double-edged sword. One the one hand, groups promote rationality by balancing out the blind spots, weaknesses, and biases of any one individual. The effective mix of skills present in the individual members of the Blockade Board illustrates this advantage. On the other hand, group dynamics introduce new sources of irrationality into decision-making such as the tendency toward groupthink.[103] As the conduct of Excomm demonstrated, however, this danger can be consciously overcome by procedural safeguards.

Another of the dangers associated with small group decision-making is when an experienced, energetic, or charismatic individual manipulates or hijacks the process to obtain the outcome he prefers.[104] Lieutenant Colonel Oliver North was able to do this to a large degree with the Restricted Interagency Group. This example reinforces the fact that the higher authority that convenes the small group to assist with the decision-making must achieve a proper balance of delegation and supervision to ensure efficient operation of the process.

Even if the leader neglects this responsibility, the small group is still an internal mechanism he created of his own volition for his own purpose. He selected the group's members to support rather than compete with his general vision, policy, and guidance. Other small groups or individuals, however, also can influence decision-making in a more competitive manner. Their agenda may differ from the decision-maker's, and their self-interest guides their actions more than a desire to help the decision-maker accomplish his objective. Oftentimes, these actors are able to exert an influence disproportionate to their numbers. The elite theory is concerned with this phenomenon.

7

The Elite Theory

Elite theory is based on the view of society as consisting of elites and masses. Although elites are few in number and masses are many, the elites control resources such as wealth, status, leadership skills, knowledge of the political process, the ability to communicate, information, and organization. They translate these resources into power, which Thomas Dye defines as "participation in the decisions that shape our lives."[1] It is this disproportionate influence of a small segment of society that sets elite theory apart from other decision-making models.

Elitism implies "that public policy does not reflect demands of 'the people' so much as it reflects the interests and values of elites."[2] Therefore, it "is vitally concerned with the identity of those individuals making foreign policy and the underlying dynamics of national power, social myth, and class interests."[3] Especially as a contrast to the rational actor model and the Weinberger Doctrine that were more common components of use of force decision-making during the Cold War, elite theory is useful in explaining the US interventions of the early post–Cold War period in Haiti, Somalia, and Kosovo.

Example 1: The Influence of the Congressional Black Caucus in the Decision to Intervene in Haiti, 1993 and 1994

Both presidents George Bush and Bill Clinton grappled with a deteriorating situation in Haiti. Viewing the problem largely from a rational actor perspective, Bush declined to intervene in the absence of a compelling national interest. Clinton, on the other hand, was beholden to the Congressional Black Caucus (CBC) in order to accomplish other items on his political agenda. The CBC made Haiti a special interest and used its power with Clinton to further its agenda. The result was a disproportionate influence of an elite group in a national decision to intervene.

Jean-Bertrand Aristide was elected president of Haiti in 1990. He was a Catholic priest who had risen to prominence in 1986 as a vocal critic of Jean-Claude Duvalier, the former president who had resigned in 1986 and left the country in chaos. Aristide's popularity with the poor and his advocacy of his own loosely defined version of socialism alienated him from the Haitian elites, who viewed him as a threat to their status quo power. On September 30, 1991, Aristide was ousted from office in a military coup led by Lieutenant General Raoul Cedras, Aristide's erstwhile hand-picked chief of staff. [4]

Even before the coup, large amounts of international aid to Haiti already had been suspended because of its governmental instability. With the government unable to pay its rank and file troops, many soldiers began resorting to armed bank robberies and home invasions as a source of income. Seeing the crime wave as a preferable option to a mutiny within the unpaid ranks, junta leaders did little to suppress the activity. The marauders were given the folkloric term *zenglendo*, which connoted "at a time when the Haitian people needed to trust state authority the most, the army had transgressed the public confidence ... and had turned on the populace in new and treacherous fashion." By the time of the Cedras junta, the army had become the "main obstacle to law and order," with the *zenglendo* often carrying out their crimes in loosely organized gangs behind the special protection of the military. Amid such chaos, thousands of Haitian "boat people" fled across the Caribbean for Florida.[5]

Haiti's plight soon attracted international attention, but in February 1993, Cedras deflected the attempt of United Nations negotiator Dante Caputo to arrange for the deployment of international human rights observers to monitor conditions in the country. The situation in Haiti was one of several international events that caught the US at an awkward policy transition between the Bush and Clinton Administrations while it struggled to define its role in the post–Cold War world. Both presidents faced pressure from domestic constituencies as they grappled with the increasing numbers of refugees fleeing Haiti by boat for Florida. While President Bush relied largely on the traditional rational actor model to guide his policy, President Clinton would become more susceptible to the influences of elites.

The Bush Administration's initial policy in dealing with the mounting refugee crisis following the ouster of Aristide was to have the Coast Guard intercept the boats and take the Haitians to a makeshift camp at the US Navy base at Guantanamo Bay in Cuba. There extensive asylum interviews were conducted to determine those whose claims were based on human rights and political issues versus purely economic ones. By the end of February 1992, almost thirty-three percent of the Haitians who reached the base at Guantanamo were permitted to apply for asylum.

In May, however, the Bush Administration responded to domestic concerns about illegal immigration and decided to close the Guantanamo camp and begin escorting detained boats back to Haiti without any asylum review at all. In explaining the new policy, Richard Boucher of the State Department stated, "It was increasingly clear that [the Guantanamo camp] was acting as a magnet and causing more Haitians to get on boats in the hopes of getting there."[6] By June, the policy change had effectively curtailed the exodus.[7]

As a presidential candidate, Clinton decried Bush's "cruel policy of returning Haitian refugees to a brutal dictatorship without an asylum hearing," and as president-elect he insisted the US "should have a process in which these Haitians get a chance to make their case." However, coming to fear such pronouncements might unleash a new wave of refugees, Clinton stated in early January 1993 that he would reverse his position and maintain the Bush policy. "The practice of returning those who flee Haiti by boat will continue, for the time being, after I become President," he explained in a broadcast to Haiti and to Haitians in the United States. "Leaving by boat is not the route to freedom."[8]

Such a stance no doubt pleased Democratic Governor of Florida Lawton Chiles, who was facing a tough reelection challenge in 1994.[9]

President Clinton might have been able to maintain such a policy during the Cold War era of national security consensus, but by the time of his presidency, "foreign affairs agendas—that is, sets of issues relevant to foreign policy with which governments are concerned—[had] become larger and more diverse."[10] The rational actor method of decision-making, which assumes a unitary state "viewed as calculating and responding to external events as if it were a single entity," was also under increasing challenge from other models.[11] Among the alternatives was the elite theory. Under this construct, "foreign policy is formulated as a response to demands generated from the economic and political system. But not all demands receive equal attention, and those that receive the most attention serve the interests of only a small sector of society."[12] In the case of decision-making about Haiti, this "small sector of society" was the Congressional Black Caucus.

The CBC's interest in Haiti was natural since the country's history is inexorably connected to issues of race. Once a French sugar colony, Haiti became an independent nation on January 1, 1804, after a remarkable revolution led in part by Toussant Louverture. Although the revolution ended slavery in Haiti, it solidified racial inequality in that rule now predominantly rested with mulattos who have been free before the revolution and desired to perpetuate the plantation economy at the expense of black laborers.

Haiti struggled to develop as a nation amid much political instability and domestic turmoil. The chaos prompted a deployment of US Marines to restore order in 1915. Among the consequences of this intervention was the imposition of the Jim Crow standards of the American South on the Haitian population. According to one study, "racism had a poisonous influence on what was already a dubious American presence."[13] Such a past relationship helped make Haiti an important cause for the CBC.

Unlike President Bush, President Clinton could not keep his distance from this powerful interest group. The CBC's numbers had climbed from twenty-six to forty after the 1992 election, and Clinton needed its members' votes for his domestic agenda.[14] According to one observer, "With a liberal agenda and a new willingness to use obstructionist tactics to see it enacted, the liberal body of black legislators exercises considerable influence at the White House these days. It's a marked contrast from the eighties when Ronald Reagan regularly turned down their invitations to meet."[15] As predicted by elite theory, the CBC was able to use this new leverage to determine what issues received attention and therefore direct the government to respond.[16] In late 1992, the CBC urged the incoming Clinton Administration "to focus its intention, not only on the refugee issue, but to attack its cause by demonstrating its unequivocal support for the restoration of democratic government in Haiti" and the return of Aristide. The CBC also asked Clinton to implement a policy of "equitable treatment of refugees regardless of color."[17]

The result of such influences was that President Clinton ultimately became "whipsawed by competing domestic pressures" that left "the administration's credibility [on Haiti] ... at stake."[18] Indeed, an article in *Newsweek* would later describe Clinton's policy toward Haiti as having "had the consistency of a Nerf ball."[19] In such an environment, there was little national consensus on what course to pursue. Certainly, any significant intervention, especially a risky military one, lacked legitimacy with the American public.

Nonetheless, the CBC kept the pressure on President Clinton, and in March 1993 he declared his intention to restore Aristide to power and help rebuild Haiti's economy. The next month, Cedras acquiesced to resign in exchange for amnesty for himself, his family, and his staff. Aristide agreed to the conditions, and UN Special Envoy Caputo returned to Haiti to begin facilitating the process.[20]

Upon his arrival, however, Caputo was met by renewed resistance from Cedras. Seemingly unconvinced that the international community was prepared to act forcefully, Cedras began "playing a game, attempting to deflect increased economic sanctions by agreeing to vacate power. When pressured to leave, however, he would renege on any agreement."[21] Only after the UN Security Council took the drastic measure on June 16 of voting to impose a ban on petroleum sales to Haiti and freeze financial assets of key Haitian authorities did Cedras seem to take notice. Just four days after this United Nations Security Council Resolution (UNSCR) 841 went into effect, Cedras and Aristide met separately with mediators at Governor's Island, New York to work out a plan to return Aristide to power.[22]

The resulting Governor's Island Accord was signed on July 3 and contained provisions for amnesty for those Haitians who had participated in the 1991 coup, the lifting of sanctions imposed by UNSCR 841, Cedras's retirement, and Aristide's return to Haiti on October 30, 1993. It was a deeply flawed document, which according to one Haiti expert left the Haitian military with "so much to lose and so little to gain."[23] Shortly after the agreement was signed, Haiti plunged into its worst period of violence since the coup. Pro-Aristide activists were routinely beaten, intimidated, or arrested. Corpses were deposited on the doorsteps of hotels where UN observers lived. Gunfire became a regular sound, and thousands of Haitians were killed or disappeared. Rather than preparing for a departure, Cedras appeared to be consolidating his power.[24]

On September 23, 1993, UNSCR 867 was passed, authorizing "the establishment and immediate dispatch of the United Nations Mission in Haiti (UNMIH)" to Haiti. The US was already make some preparations for such a measure. In August, after the signing of the Governor's Island Accord, the Joint Chiefs of Staff had directed the creation of Joint Task Force Haiti Assistance Group (JTF HAG). Colonel James Pulley, commanding the 7th Special Forces Group at Fort Bragg, North Carolina, was designated the commander. The JTF's mission was to "deploy to Haiti under United Nations operational control and conduct military training and humanitarian/civic action programs in support of Haitian democratization." [25]

In spite of this anticipation, JTF HAG was a hastily assembled group that was planned in the most cursory of manners. The implications of the Governor's Island Accord seemed to catch planners off guard, and when JTF HAG member Lieutenant Colonel Phil Baker arrived at the planning cell, he found that

> everything was in chaos. Planners from all services were thrown together trying to figure out what they were doing without much organization. Lots of people were just doing what they thought they needed to do; what they were comfortable with whether or not it had anything to do with the plan. Everybody at least looked busy. In the middle of the chaos was a Marine lieutenant colonel under a lot of pressure trying to produce an operations order. I remember that chairs were scarce; if you left yours for even a second, someone stole it.[26]

Joint Task Force HAG was certainly not the product of the military's familiar decision-making process. Instead, Walter Kretchik describes it as "an ad hoc organization whose personnel ranged from various subject-matter experts on Haiti to officers who knew nothing about the country and its problems." "Many assigned to the JTF," according to Kretchik, "had little idea of what they were expected to do."[27] The decision to intervene in Haiti bore little resemblance to the rational actor model and neither would the planning to implement that decision.

Nonetheless, pursuant to UNSCR 867, two US tank landing ships (LSTs) were prepared to transport JTF HAG to Haiti. The USS *Harlan County,* under Commander Marvin Butcher, left first, arriving at Port-au-Prince on October 11 with 225 UN observers. The USS *Fairfax County* was scheduled to follow later. Butcher's mission was to transport JTF HAG to Haiti and then provide berthing and life support to the embarked troops until they moved on to the dock. Once the landing was complete, the troops would come under the command of Colonel Pulley, who had flown to Haiti earlier and would meet the ship at Port-au-Prince.[28]

When the ship arrived in Port-au-Prince at 2:00 A.M., Butcher had to carefully navigate his way through a maze of vessels that were anchored around the harbor approaches in what appeared to be a deliberate attempt to impede his access to the port. Butcher finally dropped anchor at 5:00 A.M. but could not berth his ship because an old Cuban tanker was occupying his mooring. A US Coast Guard commander who was serving as an attaché at the American Embassy arrived at the pier only to report he was leaving due to gunfire. Butcher could hear shots as well, but they were not directed at him.[29]

Returning to the *Harlan County,* Commander Butcher reported the situation to his headquarters in Norfolk and directed all JTF HAG personnel to go to their rooms and wait. Several Haitian boats of assorted descriptions, some flying the Duvalier-era Tonton Macoute flag, were now circling the ship, but they dispersed when Butcher ordered crew members to man the *Harlan Country's* .50 caliber machine guns.[30]

From his vantage point on-shore, Colonel Pulley could also sense the mounting tension. At 7:00 A.M. he had seen a bus full of about forty Haitians arrive at the dock. Fueled by freely dispensed liquor, the crowd worked itself into a frenzy, firing weapons in the air and chanting anti–American slogans. Pulley saw two corpses dragged off the bus and thrown into the mob, but he took some comfort in the fact that a fourteen-foot high fence and a two-and-a-half-foot thick masonry wall separated the demonstrators from the pier. The mob was loud and unruly, but without access to the pier, Pulley felt it posed little immediate threat to the *Harlan County.*[31]

By this time several Americans, including Charge d' Affaires Huddleston, Pulley, and Dr. Bryant Freeman, an expert on Haiti from the University of Kansas, had gathered on the balcony of the Montana Hotel to observe the scene. Against Pulley's warnings, Huddleston decided to drive to the dock area in an attempt to calm the situation. Upon her arrival, she found the gate to the port locked, and her armored car was quickly surrounded by a mob of drunken Haitians who were chanting, "Remember Somalia." The protestors began beating on the car with ax handles, compelling Huddleston to reluctantly retreat. A live CNN video of the event gave the world a startling impression of the chaotic situation unfolding at Port-au-Prince.[32]

The harried peacekeepers spent a tense but quiet night, but the next morning, two Haitian twenty-five-foot Montauk gunboats, armed with .50 caliber machine guns and carrying Haitian Police and Haitian Army and Navy personnel, emerged from Admiral Killick Naval Base to the south and raced toward the *Harlan County.* Butcher ordered all guns manned and positioned snipers along the deck, instructing his men to open fire if the Haitians so much as put their hands on the triggers of their machine guns.[33]

Butcher concluded his position was untenable, and especially with the threat posed by the Haitian gunboats, he was unwilling to risk another night in the harbor. He notified his headquarters in Norfolk that he was pulling out. Within days, Pulley and his JTF HAG advance party were ordered out of Haiti. The remaining UN and OAS personnel soon followed.[34] In the words of Richard Millet, the Haitian "military and their supporters had, at little apparent cost, called the Clinton administration's bluff and won."[35]

That a crowd of thugs could repulse the world's greatest superpower is a testament to the scant popular support for military action in Haiti among most Americans. Part of the explanation for this indifference was that, as Bush Administration officials repeatedly emphasized, there was no vital American interest at stake in Haiti.[36] However, as the world moved further from the Cold War era, proponents of a foreign policy that supported not just US interests, but also US values, began gaining credence.[37]

Michael Mandelbaum believes President Clinton moved too far in this direction, citing Somalia, Bosnia, and Haiti as examples.[38] Mandelbaum derides Clinton for practicing "foreign policy as social work," but admits that of these three cases, Haiti was the "one place where an appeal to values might have generated support." According to Mandelbaum, because Haiti was "nearby, poor, weak, had once been occupied by the United States, and was populated by descendants of African slaves, the United States had reason to be concerned about its fate." He believes the provision of political and economic development to Haiti "could have been presented as a good deed in the neighborhood at manageable cost and justified by the fact that America is a rich, powerful, and generous country."[39] Proponents of an active American involvement in Haiti would have to convince the domestic audience that the US had a legitimate role to play, but Mandelbaum criticizes the Clinton Administration for not trying to make this case and forfeiting the chance to establish legitimacy and support for the intervention in the eyes of the American people.[40] Indeed, one of the main criticisms of elitism as a decision-making model is that it bypasses the popular participation that is the very soul of a democracy.[41]

In spite of the *Harlan County* debacle, the United States remained interested in Haiti, by one explanation, simply because President Clinton had made it an issue.[42] Perhaps Clinton had little choice, as he was still under pressure from a variety of sources, most notably from the CBC. Indeed one observer noted the CBC "has been a spiral of influence. The President has listened and the voices have been raised. The President has responded and the voices have been raised further."[43]

As elite theory would predict, the CBC had indeed been able to garner a disproportionate amount of access and attention. The New York *Times* reported that one lawmaker complained, "Administration officials consult more with key members of the black caucus about the [Haitian] crisis than they do with the chairmen and ranking members of House and Senate committees with jurisdiction over foreign policy or Caribbean affairs."[44] Critics would consider this bypassing of the established hierarchy as another threat that elitism poses to normal democratic procedures.

In March 1994, the CBC introduced a bill to tighten the economic embargo against Haiti, sever its commercial air links to the United States, halt the summary repatriation of Haitian refugees picked up at sea and block financial assets held in America by Haitian nationals. According to one Congressional staffer, the bill became "a blueprint for what was done in the coming months. This is what they rallied around and pushed for. And they got almost everything." Indeed, a State Department official admitted, "The basic components of the black caucus approach — the military is the problem, Aristide is the solution; we shouldn't move away from him even two inches; we should do nothing that smacks of any kind of alternative to Aristide, like work with a prime minister — all that has been adopted." [45]

"We are declaring war on a racist policy and an inhumane policy," explained Congressman Major Owens, head of the CBC Haiti Task Force.[46] On March 18, all forty CBC members signed a letter to President Clinton that argued, "The United States Haiti policy must be scrapped."[47] Aristide tapped into this sentiment, and on April 6, he also accused the administration of a "racist policy" toward Haitians. A few days later, Randall Robinson, the executive director of TransAfrica, announced he was starting a hunger strike to protest the US policy toward Haiti. Members of the CBC stood by Robinson as he made his statement. Congresswoman Maxine Waters declared, "I strongly support the courageous hunger strike by Randall Robinson. The Congressional Black Caucus has as its top priority the return of President Aristide and the restoration of democracy in Haiti." Waters argued, "We cannot continue to turn back Haitians fleeing for their lives by boat from oppression at the hands of the murderous military and political chief."[48]

Robinson's case drew significant media coverage, especially after he dramatically indicated his life was now in President Clinton's hands. Clinton was obviously affected. On April 19 he said, "I understand and respect what [Robinson's] doing, and we ought to change our policy. It hasn't worked."[49] Still Robinson kept up the pressure. After being checked into a hospital due to an irregular heart beat on May 5, Robinson chided, "The President has no moral core."[50] By the time Robinson ended his strike on May 8, Clinton had agreed to change the procedures to allow each Haitian to make a case for asylum.[51]

Among the key developments was the dismissal in May of Lawrence Pezzullo as the administration's special adviser on Haiti. Pezzullo have been one of the architects of the plan by which Aristide would govern by consensus or "power-sharing" with his opponents. The CBC objected to such a course, and had pushed for the unilateral restoration of Aristide to power.[52] Pezzullo was replaced by William Gray, a former member of the CBC. Although Gray dismissed suggestions that his appointment was based on his ties to the CBC, Clinton admitted, "We know about the fact he was in the black caucus, and we see that as very, very encouraging."[53] Likewise, in such moves, the New York *Times* concluded, "the caucus has played a key role in steering Haiti policy in the Administration."[54]

In taking the assignment, Gray retained his status as a private citizen and worked without pay. His appointment was limited to no more than 130 days, and he wasted no time in transmitting the message that he meant business. Soon after assuming his post, Gray made it very clear that "the military option is on the table."[55] Indeed, that is the direction to which the administration moved.

President Clinton addressed the nation on September 14 declaring that "beyond the human rights violations, the immigration problems, the importance of democracy, the United States also has strong interests in not letting dictators, especially in our own region, break their word to the United States and the United Nations."[56] The popular response to the speech was mixed, but a Gallup poll the next day reported fifty-six percent of those surveyed supported an invasion. It was far from an enthusiastic mandate, and one observer concluded, Clinton had "led the nation 'kicking and screaming' toward an unpopular intervention."[57] Such an assessment is consistent with elite theory's assertion that "public policy reflects not the demands of the masses, but the prevailing will of the elite."[58]

In spite of little support from the general public, pressure from the CBC could not be ignored. According to one observer, "under the circumstances, a strong US response to the Haitian crisis was one course of action that offered Clinton a way to extract him from a delicate political situation."[59] Such a conclusion is a powerful testimony to the explanatory power of the elite theory. It also points out its dangers. Congressman Gary Franks, at the time the only Republican member of the CBC, worried that in invading Haiti, the Clinton Administration had "blatantly ignored" the wishes of the American people. "A majority of the Congressional Black Caucus wanted the United States to invade Haiti, and President Clinton caved in," argued Franks. "This is the fundamental problem of allowing caucuses and special interests to have a disproportionate influence on US foreign policy."[60]

The day after the speech, President Clinton notified the Joint Chiefs of Staff that he had decided to authorize a military invasion, but before this option was executed, a last-minute negotiating team of President Jimmy Carter, General Colin Powell, and Senator Sam Nunn succeeded in negotiating a peaceful transfer of power from Cedras to Aristide. On the strength of this development, some 20,000 troops entered Haiti peacefully as part of Operation Uphold Democracy.

The lukewarm support among the American public for the Haiti intervention paled in comparison to the partisan debate in Congress. Just seven weeks after the deployment, Republicans won control of Congress and began to challenge the operation. Funding gradually decreased while legislative restrictions on its use increased. After six months, the US passed the mission to the United Nations but both "US and international forces departed before a competent administration could be created, self-sustaining democratic structures could be put in place, or lasting economic reforms could be institutionalized."[61] Because the intervention was so profoundly the product of elite theory, it lacked the widespread base of support to generate the perseverance and commitment the situation demanded.

Example 2: The Role of the Media in the Decision to Intervene in Somalia, 1992

Among elite theory's representations of the "principal institutions of power," Thomas Dye argues are corporations, banks, investment firms, media giants, foundations, and "think tanks."[62] Somalia is one of the most commonly cited examples of the media's role in making decisions about the use of force. Indeed, *TV Guide* crowed, "Somalia is an American foreign policy first: a military operation launched by the evening news."[63] Likewise, *Time* magazine reported, "it was pictures— of spectral women and withered children—that launched the rescue mission in Somalia."[64] While it should be noted that there are numerous counter-arguments that suggest such assessments of the media's role in this instance are exaggerated, qualitative analysis based on the statements of those associated with the Somalia decision-making process clearly indicates the influential role played by the media. [65] Like Haiti, the episode also illustrates the changing conditions in which decision-makers found themselves in the post–Cold War environment.

By the 1990s, decades of anarchy, drought, civil war, and banditry had reduced Somalia, a country encompassing approximately 637,540 square kilometers on the Horn of Africa, to a virtual wasteland. Some 300,000 Somalis died between November 1991 and March 1993 alone, and another 1.5 million lives were at immediate risk because of famine. Nearly 4.5 million of Somalia's 6 million people were threatened by severe malnutrition and related diseases. Another 700,000 had sought refuge in neighboring countries. To help relieve the mass starvation, the United Nations Security Council approved Resolution 751, which established a humanitarian aid mission known as United Nations Operation in Somalia (UNOSOM I) in April 1992. United Nations Operation in Somalia I's success was severely limited because Somali warlords, most notably Mohamed Farah Aideed of the Habr Gidr subclan and Ali Mahdi Mohamed of the Abgal subclan, refused full cooperation, and the limited mandate was not strong enough to compel compliance. The warlords kept the UNOSOM I troops from leaving Mogadishu Airport, and only 500 of the authorized 3,500 troops deployed. [66]

The failure of UNOSOM I quickly became apparent, and the US found itself under increasing pressure to act. Decision-makers were confronted with a situation different from the familiar metrics that had guided US military interventions during the latter years of the Cold War. Then, after the 1983 disaster in Beirut, Secretary of Defense Caspar Weinberger had developed what became known as the "Weinberger Doctrine"— strategic criteria to help guide "the painful decision that the use of military force is necessary to protect our interests or to carry out our national policy." [67] The overarching concern of the criteria was that the objective, commitment, and other conditions would be strong enough to ensure perseverance. The Weinberger criteria required the following:

- The United States should not commit forces to combat unless the vital national interests of the United States or its allies are involved.
- United States troops should only be committed wholeheartedly and with the clear intention of winning. Otherwise, troops should not be committed.
- United States combat troops should be committed only with clearly defined political and military objectives and with the capacity to accomplish those objectives.
- The relationship between the objectives and the size and composition of the forces committed should be continually reassessed and adjusted if necessary.
- United States troops should not be committed to battle without a "reasonable assurance" of the support of US public opinion and Congress.
- The commitment of US troops should be considered only as a last resort. [68]

The Weinberger Doctrine had emerged as the definitive yardstick for measuring the application of military force, but by the time of the Somalia crisis, the success of Operation Desert Storm and the collapse of the Soviet empire were changing both America's perception of its military and the nature of the threat. To many, Weinberger's strict criteria for the use of force seemed to require revision. [69] They sought new roles and missions for the US military in the globalized era.

The decision to intervene in Somalia reflected these changing thoughts on the commitment of force. In the case of Somalia, vital American national interests were not at stake. In their place were the peripheral interests of promoting American values and a favorable world order. While such interests did not meet Weinberger's high bar, it appeared that the application of military power was likely the only option that had a reasonable prospect of producing favorable results in Somalia at an acceptable cost.[70] A variety of interest groups began advocating military intervention.

Thus, rather than being the result of deliberate and methodical exposure to a criteria such as Weinberger's, the decision to intervene in Somalia was subjected to a series of domestic and international influences that were consistent with the predictions of the elite theory. The most conspicuous factor was an onslaught of media coverage that portrayed starvation conditions in Somalia and created the impression that only US intervention could save the country. The ability of the media to project such compelling imagery was a fairly recent phenomenon.

In 1980, the Cable News Network or CNN became the first television channel to provide twenty-four hour a day news coverage. With the deployment of communications satellites, CNN and other modern news outlets could now broadcast "real time" reports from anywhere on earth. Collectively, this expanded news coverage creates what many observers call the "the CNN curve" or "CNN effect" which "has come to represent the influence that this new kind of 'real-time' reporting can have—that dramatic images of starving masses, shelled populations, or dead American soldiers can induce public demands for action from elected officials."[71]

The CNN effect manifests itself in several ways concerning foreign policy decision-making, but the one most present in the case of Somalia was as an "agenda setting agent." Agenda setting involves using emotional and compelling coverage of events to reorder foreign policy priorities.[72] It is in this way that the media defines the "problem" as a "problem" and "decides what is decided."[73]

The media is capable of an agenda setting role by both "framing" and "priming." Framing holds that how the media casts an issue affects substantive judgments people make about the issue. Priming argues that the priority the media gives to an issue affects the priority people give to the issue. The result is noted by Bruce Jentleson who says, "The mass media may not be successful in telling people what to think, but the media are stunningly successful in telling their audience what to think about." Jentleson then asks, "If a tree falls in the woods and CNN doesn't cover it, did it really fall?"[74]

Numerous high level decision-makers and observers recognize the impact of the media as an agenda setter. Former Secretary of State James Baker argues that "all too often, television is what determines what is a crisis."[75] President Bill Clinton's National Security Advisor Anthony Lake explains, "We know that when the all-seeing eye of CNN finds real suffering abroad, Americans want their government to act—as they should and we should."[76] Similarly, former Secretary of Defense James Schlesinger believes, "National policy is determined by the plight of the Kurds or starvation in Somalia, as it appears on the screen," and Jessica Matthews, former Deputy Under Secretary of State for Global Affairs, states, "The process by which a particular human tragedy becomes a crisis demanding a response is less the result of a rational weighing of need or of what is remediable than it is of what gets on the nightly news."[77] In the case of Somalia, Michael Mandelbaum concludes that "televised pictures of starving people created a political clamor to feed them, which propelled the US military into action."[78]

Proponents of the media's impact on Somalia focus on the statements of those in the decision-making process. White House Press Secretary Marlin Fitzwater acknowledges this impact of the press coverage, saying, "After the election [of November 1992], the media had free time and that was when the pressure started building up.... We heard it from every corner, that something had to be done. Finally the pressure was too great ... TV tipped us over the top.... I could not stand to eat my dinner watching TV at night. It made me sick."[79] Perhaps most telling is Craig Hines's report of the media coverage's impact on President George H. W. Bush. Hines writes,

> Bush said that as he and his wife, Barbara, watched television at the White House and saw "those starving kids ... in quest of a little pitiful cup of rice," he phoned Defense Secretary Dick Cheney and Gen. Colin Powell, Chairman of the JCoS [Joint Chiefs of Staff]: "Please come over to the White House." Bush recalled telling the military leaders: "I–we–can't watch this anymore. You've got to do something."[80]

Responding to such motivations, the US won United Nations Security Council approval in December 1992 of Resolution 794, which established Unified Task Force (UNITAF), a large, US-led peace enforcement operation known as Operation Restore Hope. However, these public demands for action based on media coverage can negatively affect policy decision-making and planning. Former Secretary of State Lawrence Eagleburger lamented, "The public hears of an event now in real time, before the State Department has had time to think about it. Consequently, we find ourselves reacting before we've had time to think. This is now the way we determine foreign policy — it's driven more by the daily events reported on TV than it used to be."[81] Certainly such a condition retards rational and deliberate strategic development.

During the Cold War, realism was the dominant theoretical tradition of international relations and national security was the clear priority. In the post–Cold War absence of the Soviet threat, realism's clearly stated and preferentially ranked goals became increasingly elusive. In the absence of realism's clear monopoly on "interests," more idealistic theories that stressed "values" gained credence.[82] The media elite no doubt welcomed this shift because it was compatible with their value "to use government power to 'do good.'"[83] Consistent with this development, other policy-making models such as elite theory have increasingly challenged the rational actor model.

The negative risks of such a model are obvious. Glenn Hastedt warns that "Those outside the elite group are held to be relatively powerless, reacting to the policy initiatives of the elite rather than orchestrating them. Furthermore, public reactions are often 'orchestrated' by the elite rather than being expressions of independent thinking on policy matters."[84] Therefore if the global media is one of these elite groups that is able to drive policy, then it, rather than policy makers, is the one setting much of the policy agenda. If the media "orchestrates" this agenda for considerations such as viewership and market share rather than true policy relevance, it would be leading policy makers in a direction that is not in the nation's best interests; certainly, at least, not as those interests were previously understood and defined by realism and the rational actor model.

Often times, such orchestration leads to conspiratorial charges associated with elites, which is not necessarily the case.[85] Critics often wonder why the US intervened in Somalia rather than the equally if not greater devastated Sudan which one diplomat described as "Somalia without CNN."[86] In this case, the disparity in media coverage had much more to do with geography and logistics than any conspiratorial media preference. As Kristen Lord explains, "The famine in Somalia was concentrated in a relatively accessible area, whereas suffering in Sudan was spread over great distances, making coverage logistically difficult."[87] This information differential resulted in the typical characteristic of elite theory that "not all demands receive equal attention."[88]

The end result was that the American decision to intervene in Somalia was not based on the traditional rational actor model. Indeed, one contemporary observer described it as "more generous impulse than thought-out policy."[89] It certainly gave the appearance of being a "largely tactical decision reached to solve a current, concrete problem with little apparent concern for the longer term strategic implications."[90]

Nonetheless, UNITAF made great progress, and humanitarian agencies soon declared an end to the food emergency. By January 1993, food was getting to all areas of the country.[91] Indeed, what UNITAF found was pockets of hunger rather than the widespread starvation and universal life-threatening conditions that had been portrayed by the media and demanded action.[92] In light of these improvements, US forces began withdrawing in mid–February, and on May 4, UNOSOM II, armed with a much broader mandate that included nation-building activities, took over operations from UNITAF. With the benefit of hindsight, Colonel Kenneth Allard notes that at this point "the underlying causes of conflict in Somalia had only been postponed," and it was during UNOSOM II that they resurfaced and "exploded."[93] Overlooking such critical factors is a danger inherent in straying from the rational actor model. In this case, the flawed strategy manifested itself on October 3–4, 1993, in the disastrous Battle of Mogadishu, in which eighteen Americans died.

Although it is estimated that the Americans inflicted up to two thousand casualties on the Somalis during the Battle of Mogadishu, the American losses and the chaotic nature of the operation created a domestic outcry in the United States.[94] Almost immediately, President Clinton announced the decision to withdraw US troops by March 1994. Much of this outcome can be traced to the initial influence of elite theory in the decision to intervene.

First, the decision was heavily shaped by the special interest of the media as elite theory would predict rather than the national interest as the rational actor model would have likely demanded in the Cold War. Second, the decision was disproportionately based on the influence of the media, representing a small sector of American society, rather than the legitimacy of national consensus. As a product of elite decision-making, the US intervention in Somalia suffered in terms of its objective and its legitimacy. It reflected a value promoted by an elite segment of American society to use the military to "do good" rather than a vital interest related to national security. Also as a special interest of an elite group, the intervention never evoked a deep US national commitment and could not withstand the casualties of the Battle of Mogadishu.

Still, even in the decision to withdraw, the media maintained its influential role. Images of a dazed and bruised captured pilot, Somalis dancing joyously amid helicopter wreckage, and the beaten bodies of US soldiers being dragged through the streets, invoked outrage in the American public. The impact of the media had come full circle. In his October 7 speech, President Clinton began by reminding Americans, "A year ago, we all watched with horror as Somali children and their families lay dying by the tens of thousands—dying the slow, agonizing death of starvation." After the Battle of Mogadishu, however, the media was presenting different images which Clinton addressed saying, "This past weekend we all reacted with anger and horror as an armed Somali gang desecrated the bodies of our American soldiers and displayed a captured American pilot."[95] Likewise, Chairman of the Joint Chiefs of Staff General Colin Powell explained, "We have been drawn to [Somalia] by television images; now we are being repelled by them."[96] In light of such testimony, elite theory seems to offer a viable explanation of the role the media played in influencing the US policy to first enter and then leave Somalia.

Example 3: The Agenda of Secretary of State Madeline Albright in the Decision to Intervene in Kosovo, 1999

Kosovo was another humanitarian crisis that the United States faced in the early post–Cold War years. While elite groups such as the Albanian diaspora and the media influenced the process, the main driver of the US decision to intervene was Secretary of State Madeline Albright. Albright's role in this process reflects the emphasis elite theory places on the identity of decision-makers.

After World War II, Kosovo became an autonomous province of Serbia in the Republic of Yugoslavia. It was a poor part of the country with some deposits of a variety of minerals and metals but little infrastructure or capability to realize its mining potential. Instead, what brought attention to Kosovo was its demographics. Among its two million people, the vast majority were Albanian, with Serbs compromising just ten percent of the population. That had not always been the case. During the medieval period, Kosovo became the center of the Serbian Empire and home of many important Serb religious sites, including many architecturally significant Serbian Orthodox monasteries. Thus the Serbs suffered a devastating loss when Ottoman forces won the battle of *Kossovo Polje* or the Field of the Black Birds in 1389. Five centuries of Ottoman rule followed, bringing large numbers of Turks and Albanians to Kosovo. By the end of the nineteenth century, Albanians replaced the Serbs as the dominant ethnic group in Kosovo. Serbia did not regain control over Kosovo until the First Balkan War of 1912.

Against this backdrop, Kosovo became a battleground for both Serb and Albanian nationalism. In 1987, Slobodan Milošević, a rising Serbian Communist Party leader, made a speech at the anniversary of the defeat at *Kossovo Polje* and, pointing his finger off in the distance, vowed, "They'll never do this to you again. Never again will anyone defeat you." He rode this wave of nationalism to election as the President of Serbia in 1989. Constitutionally limited to two terms in this role, Milošević then became President of the Federal Republic of Yugoslavia in 1997. In so doing, he became "the only Eastern European Communist leader in the late 1980s who managed to save himself and his party from collapse [and he] did so by making a direct appeal to racial hatred."[97]

Under Milošević's leadership, Serbia passed a new constitution in 1989 that revoked Kosovo's status as an autonomous province of Serbia and instituted a series of repressive measures against Kosovar Albanians. Kosovo's Albanian leaders responded in 1991 by organizing a referendum that declared Kosovo independent. An unofficial "shadow government" led by Ibrahim Rugova attempted to gain international assistance and recognition of an independent Kosovo. Other Albanians who were dissatisfied with Rugova's passive strategy created the *Ushtria Clirimtare e Kosoves* (UCK or Kosovo Liberation Army — KLA) and launched an insurgency in the 1990s.

In 1998, Serbian military, police, and paramilitary forces began an intensive counterinsurgency campaign that became an attempt at ethnic cleansing. The US State Department estimates some ten thousand Kosovar Albanians were killed by Serb forces, and one and a half million were forcibly expelled from their homes.[98] Many of those evicted fled to neighboring Macedonia and Albania, and a refugee crisis soon threatened to overwhelm these nations' meager resources.[99]

As President Bill Clinton wrestled with the situation in Kosovo, he turned to three primary advisors: National Security Adviser Sandy Berger, Secretary of Defense William Cohen, and Secretary of State Madeleine Albright.[100] Of this small group, Albright was clearly the dominant figure. Steven Redd reports a LEXIS–NEXIS keyword search of major media sources yielded twelve hits for "Cohen" and "Kosovo"; twenty-eight for "Berger" and "Kosovo"; and a whopping 291 hits for "Albright" and "Kosovo" for the time period January 1 through April 1, 1999.[101] Personally distracted by the Monica Lewinsky debacle, this was a time in which Clinton was particularly susceptible to the influence of advisors, and Albright seized the opportunity.[102]

In so doing, Albright represented a singularly aggressive voice among her fellows. Both Cohen and Berger had come out against military action early in the debate. Berger even went so far as to counsel against merely threatening to use force for fear of possible negative political consequences. As the situation worsened, Berger modified his stance but remained adamant against the use of ground troops.[103] Cohen also could see many causes for concern. He explained, "The notion of becoming involved in Kosovo certainly presented a number of challenges in terms of readiness of resources, the manpower that would be involved, the commitment of our troops for what length of time, and how we pay for it." What is more, Cohen recalled, "There was great reluctance on the part of most members of Congress to commit American forces, even on a peacekeeping mission."[104] In addition to these concerns, there were also competing domestic considerations. David Bergen complained, "This was to be the year that Washington agreed on an overhaul of Medicare, bipartisan reform of Social Security, and a new federal approach to education…. Then came Kosovo—and boom, the domestic agenda is going up in smoke."[105]

Secretary Albright did not share such reservations. Indeed, Michael Hirsh argues, "Again and again, she has tried to pull her boss, Bill Clinton—who is nothing if not equivocating—in a more aggressive direction." Hirsh added, however, that Albright would then "look over her shoulder and find he's not there."[106] Still, she resolutely endeavored to steer policy in her preferred direction.

What brought new urgency to the Kosovo crisis was the January 16, 1999, discovery of the bodies of forty-five peasant farmers and their children who were massacred in the village of Racak. According to Albright, Racak "was a galvanizing event…. Despite all the efforts, something as terrible as Racak could happen. It energized all of us to say that this requires a larger plan, and a steady application of military planning for an air campaign."[107] Although Cohen also considers the Racak massacre "galvanizing," he notes, "It didn't change my thinking. I felt that military force should be the absolute last resort. Everything else has to fail before you turn to the military."[108]

While Cohen clung to the traditional conservative use of force philosophy of the Weinberger Doctrine, Albright was willing to be much more aggressive. Just three days after the bodies were found, she presented a new plan at a White House meeting. The plan continued the threat of air strikes if Milošević did not agree to a peace plan, but it went on to demand that he accept NATO troops in Kosovo to enforce a deal under which Milošević would withdraw his security forces and grant Kosovo broad autonomy. Berger,

Cohen, and General Hugh Shelton, Chairman of the Joint Chiefs of Staff, opposed Albright's plan, but the Secretary of State persisted. She eventually won enough support to send her idea to President Clinton who accepted it.[109] Her role in this process was singular. According to Tom Carver, "By all accounts, it was Madeleine Albright who convinced Clinton, against the better judgment of the Pentagon, that [Milošević] would back down after a little light bombing."[110]

Redd concludes that it was at this point that Albright "began to assert her greatest influence over the foreign-policy-making process."[111] She summoned the Serbs and the UCK to the French chateau of Rambouillet and presented them with her new plan. She was a powerful and omnipresent force at the negotiations. Department of State Spokesman James Rubin reported:

> Secretary Albright is having a one-on-one meeting with [Serbian] President [Milan] Milutinovic in her hold room. She has held meetings with him with the other Ministers. She has met with the Ministers as a group. She has addressed the entire Kosovo Albanian delegation. She has met privately with three or four of the key Kosovo Albanian leaders and their work is very intense.[112]

Albright demanded that both sides sign the agreement, and then threatened American and NATO military reprisals if the Serbs refused. Pulling no punches, she advised Milošević that it was time to "wake up and smell the coffee," and at one point she warned, "It's all stick now." Even her European counterparts recognized her influence, with one diplomat observing that "she has massive clout — she's the one who can say to the Serbs, 'sign this, or we'll bomb the hell out of you.'"[113] In a last ditch effort to reach an agreement, President Clinton noted Albright "dispatched Ambassador Dick Holbrooke to Belgrade to talk to President Milošević one last time." Redd cites this as "further evidence that, among President Clinton's many advisers, Secretary Albright was the lead on Kosovo foreign policy and that she had the greatest influence on the president's decisions."[114] Indeed, Redd concludes, "Albright was the driving force behind the decision to finally confront Milošević, and to use air strikes against the Serbs."[115] Walter Isaacson adds, "more than anyone else, [Albright] embodies the foreign policy vision that pushed these men into this war. And she is the one most responsible for holding the allies — and the Administration — firm in pursuit of victory."[116] Ultimately, a variety of sources went so far as to declare the US involvement in Kosovo to be "Madeleine's War."[117]

Albright was no stranger to the Kosovo crisis, having worked issues relating to the former Yugoslavia in the early 1990s as the US ambassador to the United Nations. "I had been very concerned about Kosovo for some time," she explained. "The whole Balkan crisis began in 1989, when Milošević basically took away Kosovo's autonomy."[118] She was a staunch critic of British and French reluctance to do anything serious against the Serbs in Bosnia from 1992 to 1994 and was adamant that more decisive action would be taken in Kosovo. As early as March 1998, she declared, "we are not going to stand by and watch the Serbian authorities do in Kosovo what they can no longer get away with doing in Bosnia.... The time to stop the killing is now, before it spreads."[119]

Elite theory places great emphasis on the identity of those individuals making foreign policy, and for many observers, the roots of Albright's attachment to the developing crisis ran much deeper than even Bosnia. Albright was born in Prague as the daughter of a Czech diplomat. She had personal experience with dictatorship and terror as her family fled from the approach of Hitler's army. Her Jewish grandparents fell victim to the Holocaust. After the war, Albright's family returned to Czechoslovakia only to leave again to avoid communism. It was this move that brought the family to the United States. The result of this experience left Albright with a "highly personal approach to foreign policy," and many observers "believed that her childhood experience colors her world view and, by extension, America's foreign policy."[120] "Madeleine Albright, more than anyone else in this administration, is driven by her own biography," explained one senior US diplomat.[121] Albright would seem to agree, saying, "Identity is a complex compilation of influences and experiences—past and present. I have always felt that my life has been strengthened and enriched by my heritage and my past. And I have always felt that my life story is also the story of the evil of totalitarianism and the turbulence of twentieth-century Europe."[122]

Elite theory would see a distinct connection between this personal experience and Secretary Albright's decision-making. To that end, Walter Isaacson concludes that Kosovo was, for someone like Albright, "who had to flee Hitler, then Stalin, as a child, a very personal mission."[123] He argues the experience left her with "an instinctive antipathy toward thugs"—a category in which Albright would no doubt place Milošević.[124] Even President Clinton acknowledged the impact of Albright's frame of reference on her handling of the Kosovo crisis, saying, "Secretary Albright, thank you for being able to redeem the lessons of your life story by standing up for the freedom of the people in the Balkans."[125]

The result of these efforts was Operation Allied Force which lasted seventy-eight days from March 24 to June 20, 1999, and involved the US and thirteen NATO allies. In order to avoid casualties, it was entirely an air campaign, and indeed, it concluded without a single combat fatality to NATO forces.[126] Over 38,000 combat sorties were flown in around-the-clock operations that pounded military, command and control, and communications and transportation infrastructure targets.[127] In the wake of this destruction, Milošević consented to a Military Technical Agreement with NATO which provided for the withdrawal of Yugoslav forces from Kosovo and the entry of a NATO-led international security force called the Kosovo Force or KFOR. Under a mandate provided by United Nations Security Council Resolution 1244, the first elements of KFOR entered Kosovo on June 12. By June 20, all Serb forces had withdrawn from Kosovo. But while Operation Allied Force was a clear military success, a CBS poll found that only fifty-one percent of Americans supported the bombing campaign.[128]

The ambivalent response of so many Americans reflected the fact that "while there were strong reasons to intervene, targeting thousands of bombs on a murderous, tyrannical regime, our vital national interests [were] never at stake."[129] This condition created serious problems with strategic objectives. The crisis in Kosovo placed the United States in the difficult position of meeting a short-term objective of stopping Serb violence while at the same time facilitating the long-term objective of not encouraging independence for Kosovo or becoming tied to another long-term and potentially dangerous peacekeeping and nationbuilding operation. These short and long-term objectives were largely mutually exclusive.

Much of historic US policy with regard to supporting insurgencies and counterinsurgencies seems to have been shaped by the assumption that "the enemy of my enemy is my friend." Sometimes this short-term expedient backfires over the long-term. A good example is the US support of the mujaheddin against the Soviets in Afghanistan during the 1980s. At the time, the US supported the mujaheddin as an ally against a common enemy because they were willing and able to provide capabilities and a level of engagement that the US was not. Since then, the US has had to live with the mujaheddin's connections to radical Islamic fundamentalism and international terrorism.[130]

In Kosovo, the US entered into a similar marriage of convenience with the UCK.[131] The US needed someone who would fight Milošević on the ground when the US had the will to do it only from the air, and needed someone to cast in the role of heroic freedom fighter that would rally disparate domestic support. The US saw the UCK as fulfilling these needs. In so doing, policy-makers overlooked several key facts that had longer-term implications. Key among these was the fact that the UCK's agenda for independence in Kosovo was contrary to that of the US at the time of Operation Allied Force.

On September 24, 1999, State Department spokesman James Rubin reiterated, "We have always said we do not support independence for Kosovo, and we do not support independence for Kosovo now."[132] Yet in supporting the UCK, the US found itself in an awkward relationship with a partner who was "uncompromising in its quest for an independent Kosovo now and a Greater Albania later."[133] Indeed, in 2008 Kosovo unilaterally declared its independence, and the US recognized the declaration. Kosovo is hardly a self-sustaining nation, however, and at the start of 2013, KFOR still consisted of 5,565 troops, of which 760 were American.[134]

Although the US has been unable to extricate its military from Kosovo as it did in Haiti, the interventions are similar in that they represent the difficulty of satisfying national interests and objectives in predominantly humanitarian situations. Carl Hodge warns "national governments have become ... inclined to regard military force in the name of human rights as an addition to their toolkit for short-term political and diplomatic contingencies that offer little of lasting value to international peace."[135] He offers that, "The lesson of Kosovo, surely, is ... that intervention is likely to be effective and perceived as legitimate *only* when it is married to credible self-interest. The absence of an abiding Western interest in the future of Kosovo was reflected in the qualified nature of NATO's action at every level, from diplomacy to restricted bombing."[136] The challenge Hodge identifies is "establishing a clarity of purpose that is both humanitarian and credibly self-interested."[137] It was a combination that eluded the US in Kosovo.

Elite theory would predict such a deviation from national interest and the accompanying lack of consensus and solidarity. According to one observer, crises like Kosovo "are too small-time to arouse a national consensus to go to war, yet too annoying to be ignored, especially when they're always popping up on CNN."[138] Such situations are conducive for elites like Albright to take the initiative and give direction to an otherwise murky situation.

Albright was well-aware that her approach had served to isolate her, according to Operation Allied Force's NATO commander General Wesley Clark. He reports Albright confided to him, "Well, Wes, it's up to you. I've done my best, and they've called it my war, and they've turned on me."[139] Decision-makers under elite theory may be successful in accomplishing their agenda, but they do so with a cost. By garnering more than their share of influence, they can alienate others. By deviating from the rational actor model, they can elevate special interest above national interest.

But while elites may end up somewhat isolated, their grip on power also makes them somewhat insulated. They are simply better equipped to work the political process than non-elites, and their commitment to their agenda gives them a focus and purpose they can use to their advantage. As Thomas Dye explains, "active elites are subject to relatively little direct influence from the apathetic masses."[140] Certainly in pressing her case for an intervention in Kosovo, Secretary of State Albright was able to use passion and action to overcome indifference and negativity.

Utility of the Elite Theory

Although not restricted to applicability in the post–Cold war era, elite theory certainly is useful in explaining decisions made in such periods in which there is a fracturing of opinion concerning the use of force. With the end of the Cold War, the familiar consensus on the national interest was replaced by less-defined values. As various groups asserted their particular value, elite groups and individuals were able to leverage their power, position, and perspective to advance their agenda.[141] But while elite theory is useful in explaining decisions involving the use of force in cases such as Haiti, Somalia, and Kosovo, many observers are alarmed by the seeming affront to democratic principles that such disproportionate influence implies. For them, the pluralist model represents much more of the sharing of power that should characterize decision-making in a liberal democracy like the United States.[142]

8

The Pluralist Model

Both elite theory and pluralism agree that power in society is widely dispersed, but while elite theory sees the division between elites and the masses, pluralism sees the competition as being between leadership groups. There are multiple leadership groups which all have different types of power and different abilities to influence different decisions. The structure of power is therefore "polyarchal," requiring the groups to bargain and compromise over key decisions.[1]

In this process, power resources can be substituted for one another. For example, large numbers can offset wealth; superior leadership can offset large numbers; and dedication can overcome poor leadership. The result is a reinforcement of the democratic process, because countervailing centers of power each check the ambitions of others.[2]

Glenn Hastedt identifies six common themes associated with pluralism:

1. Power in society is fragmented and diffused.
2. Many groups in society have power to participate in policy making.
3. No one group is powerful enough to dictate policy.
4. An equilibrium among groups is the natural state of affairs.
5. Policy is the product of bargaining between groups and reflects the interests of the dominant groups(s).
6. The government acts as an umpire, supervising the competition and sometimes compelling a settlement.[3]

Pluralism is useful in explaining conscription in the Confederate Army, the US reflagging of Kuwaiti tankers in the Iran-Iraq War, and the commitment of the Implementation Force to Bosnia in 1995.

Example 1: Conscription
in the Confederate Army, 1861–1864

In its ideal form, pluralism as a decision-making model is often touted as being very compatible with democratic principles. While, like elite theory, pluralism recognizes the presence and power of interest groups, it believes each individual citizen is represented by one or more of those groups. The result is that "the struggle for influence over policy resembles a multisided tug of war in which each interest group attempts to pull the government in its own direction while resisting the tugs of others. From this contest, policy emerges."[4]

While such a model may reflect an equitable representation of the society and prevent the domination of a single interest, its power-sharing does not necessary produce the optimal solution to the problem from the standpoint of national security. Pluralism does not feature the assumption of a unitary state as in the rational actor model. While a common decision may emerge from the pluralist model, it is as much the product of a bottoms-up as a top-down process, and therefore is profoundly shaped by those decentralized interests.

The fledgling Confederate nation suffered from this phenomenon in a variety of ways, including its efforts at conscription. The various interest groups that represented the Confederate citizenry were the military, the state governments and the proponents of states' rights, slaveholders, certain professions and industries, and the potential draftee and his family. As pluralism would predict, President Jefferson Davis and the central Confederate government acted as an umpire in this competition, trying to craft a conscription policy that reflected the fragmented and diffused power in the society while still meeting the national security need. This was no small challenge for a new nation in the midst of a war, and the effort to account for the disparate interests has led some observers to believe the tombstone of the Confederacy should read "Died of Democracy."[5]

The military's interest in conscription came from the obvious need to build an army from scratch. The "Army of the Confederate States of America" was established by the Provisional Congress on March 6, 1861, but never really came into existence. The organization that fought the Civil War on behalf of the Confederacy was the volunteer or Provisional Army established by Congress on February 28 and March 6. Under these acts, President Davis assumed control of military operations and received state forces and 100,000 volunteers for a one-year term of service. Under this authority, Davis had called for 82,000 men by April. On May 8, Congress voted to extend enlistments for the duration of the war. While the Battle of First Manassas on July 21 was a Confederate victory, it also dispelled any notions that the war would end quickly. On August 8, Congress authorized 400,000 additional volunteers for periods from one to three years. All of these troops entered the Army via the states. While the Confederacy had started the difficult process of building an army, it had not yet produced one with a sufficient, predictable, and reliable personnel strength.

Even before he became Secretary of War on March 22, 1862, George Randolph had been considering how to address the manpower problem the Confederacy would face when its initial enlistments expired. On March 28, President Davis acted on Randolph's recommendations and requested the Confederate Congress pass the first military draft on the North American continent. Although the measure posed a challenge to the traditional Southern characteristics of state's rights and individualism, Congress approved the legislation on April 16. This marked the first time soldiers directly entered the Provisional Army rather than passing first through the states. Under the act's provisions, all white males between eighteen and thirty-five years of age were obligated to three years' service, or less if the war ended sooner.[6]

A second conscription law in September raised the upper age limit to forty-five. Those already in the service would remain in it, and the twelve-month volunteers received a 60-day furlough. From a military perspective, the conscription acts were critical in that they kept veterans in the Army at a decisive time. Thus, Richard Beringer and his colleagues credit the measures with "the salvation of the Confederate Army" after the battlefield disasters of early 1862.[7] Unfortunately for the Confederate government, while conscription served its military purpose, such acts, deemed as coercive by other interest groups within the Confederacy, "were subjected to sabotage on every hand."[8]

While conscription was "a significant step toward improving Confederate military strength," it was an equally dramatic move "away from the traditions of state rights and individualism characteristic of the ante-bellum South."[9] For many, it represented the same centralization of power and imposition of uniformity that had helped spark secession.[10] "At one fell swoop," Governor Joseph Brown of Georgia declared, conscription "strikes down the sovereignty of the State, triumphs upon the constitutional rights and personal liberty of the citizens, and arms the President with imperial power." It reduced "free-born citizens of the respective States," Brown continued, to "vassals of the central power."[11] Because state and individual interests were represented by powerful groups, the conscription policy officially and unofficially deviated from its purest form. These accommodations show the impact of pluralism on both the decision-making process and the policy's implementation.

Many Southerners felt conscription "represented the first step toward military despotism." They understood that, of necessity, the strength of the military would increase during wartime, and they demanded a counterbalance to this growth. According to Marc Kruman, North Carolinians, led by Governor Zebulon Vance, in particular felt a strong state government was a critical safeguard to the people's liberties and would serve as a buffer to central power.[12]

Conscription was a major expansion of central authority, in that the national government, rather than the states, was given responsibility for recruiting and organizing units. As a result, many state's rights advocates objected more to the form conscription took rather than the idea itself. Senator James Orr of South Carolina, for example, proposed a policy by which the president would establish a quota for each state, deduct from that quota the number of the state's citizens currently serving, and then requisition the state's governor to provide the remainder. If the governor failed to do so, the president would call out the state militia. While Orr's proposal was not adopted, the original conscription act did bow to state's rights by requiring that draftees would "be assigned to companies from the State from which they respectively come."[13]

Concerning conscription's perceived infringement on individualism, the Confederate government made a major concession in giving soldiers the privilege of electing officers at the company level. Since units were often formed locally, relatives, neighbors, and friends commonly served in the same company, making the elections very personal affairs. Campaigns were conducted with all the promises and machinations of any political process. Such a system disadvantaged otherwise competent officers who the men considered overly strict. The result was that in many cases, popular rather than competent officers were placed in command.[14]

Substitution was another means of mitigating resistance to the draft. As early as the fall of 1861, the Confederacy began releasing volunteers from the Army if they could furnish an able-bodied replacement to serve in their stead, and the practice of substitution was codified in the first Conscription Act. Under these procedures, a conscripted man could report to the camp of instruction with a substitute, and, if the substitute was found medically qualified and not subject to military service of his own, the originally drafted soldier could return home, leaving the substitute to fulfill his obligation.[15]

Many saw substitution as unfairly favoring Southerners of means, and protests against the practice increased as the war progressed. Soldiers returning home on leave from the front lines found it demoralizing to see able-bodied men who had avoided service. It was seen by many as evidence that the Confederacy was engaged in "a rich man's war and poor man's fight." Nonetheless, the practice grew and became somewhat of a cottage industry, complete with its own brokerages. Newspapers carried advertisements offering substitutes for prices ranging from $500 in 1862 to thousands of dollars toward the end of 1863. In November 1863, Secretary of War Seddon offered the conservative estimate that some 50,000 men had obtained substitutes.[16]

Inevitably, the system became adulterated by fraud and corruption. The provision allowing only one substitute per month in each company was frequently violated. Some substitutes subsequently deserted, only to offer their services again under a different name. Unscrupulous examiners allowed unfit substitutes to be accepted. Most substitutes were motivated solely by mercenary gain and had little devotion to the cause. Throughout the war, the system defied reform, and in early 1864, Congress abolished it altogether.

Pure conscription was further weakened on April 21, 1862, when procedures were added to cover those exempted from service based on the possession of special skills or the holding of important positions that made the individual more valuable in a civil than a military capacity. Classes of exemption included national and state officers, railroads workers, druggists, professors, schoolteachers, miners, ministers, pilots, nurses, and iron-furnace and foundry employees. The War Department was authorized to grant other exemptions as necessary. Still others were exempted after failing medical examinations.[17]

The substitute and exemption systems created opportunities for other interest groups to challenge central power. Abuses of the medical exemption were so outrageous that Major General James Kemper reported to General Robert E. Lee his opinion that of the 19,000 men in Virginia that had been medically exempted by district boards, 10,000 would be found fit for service if subjected to "a rigorous re-examination."[18] Other challenges manifested themselves under the auspices of the same doctrine of state's rights that the Confederate states had used to justify secession. Governors Brown of Georgia and Vance of North Carolina are the two most notorious examples. Both men resisted President Davis's efforts to centralize power at the national level in general and conscription in particular. Such tactics included creating superfluous state offices to facilitate exemptions and demanding immunity from the draft for such petty officials as justices of the peace by claiming they were essential to the discharge of governmental functions. They claimed conscription was unconstitutional, and some state justices, notably Richmond Pearson of North Carolina, facilitated the resistance by issuing writs of *habeas corpus* on behalf of those held in violation of draft laws. Governors exempted local defense troops, depriving the Confederate Army of thousands of militiamen of conscription age.[19]

The fact that state's rights advocates viewed conscription from their own limited perspective is an example of Theodore Lowi's contention that "no individual interest group can be expected to take fullest account of the consequences of its own claims."[20] State governors and others who considered conscription only from their narrow local perspective had a luxury that President Davis, with his broad national security responsibilities, did not. Thus from his provincial point of view, Brown could with clear conscious declare conscription "unnecessary as to Georgia."[21]

Brown used his power to exempt more men than any other governor. His resistance was so profound that in November 1862 President Davis sent former interim Secretary of State William Browne on what historian Emory Thomas calls "a diplomatic mission to seek the compliance of the Georgia Governor." The whole concept in general and Thomas's word choice in particular serve to vividly illustrate the fragmented power in the Confederacy. The idea of a president dispatching an official from the department normally charged with foreign relations on a "diplomatic mission" to a state governor to "seek" compliance with a national law is a testimony to Brown's power and Davis's deference to it. Even after the Supreme Court in Georgia upheld the draft unanimously, Brown continued to resist it and withheld troops whenever possible.[22]

Indeed, the judicial system provided significant overall support for conscription. President Davis challenged recalcitrant governors to address their concerns not by defying Congress, but by appealing to the courts. "The Confederate courts, as well as those of the State," he advised South Carolina's Governor Francis Pickens, "possess ample powers for the redress of grievances whether inflicted by legislation, or Executive usurpation." Confident he would be supported by the courts, President Davis put the governors in the awkward position of being embarrassed if they defied a law that was subsequently upheld. Davis's appeal to the power of the courts to influence state decision-making is a practical example of the pluralist model at work, and his "strong letter" to Pickens "was enough to do the trick" to facilitate conscription in South Carolina.[23]

Although in some cases Governor Brown supported national efforts, such as by encouraging Georgia farmers to reduce their cotton crops and instead grow food for the Confederate Army, his overall frustration with President Davis led to Brown's eventual anti–Administration alliance with fellow Georgian, Vice President Alexander Stephens. As Stephens became increasingly disenchanting with what he saw as Davis's nationalist bent, he spent less and less time in Richmond and became less and less involved in the responsibilities of his office.[24] A situation in which power is so fragmented that even the president and vice president operate at cross-purposes is clearly pluralism run amuck.

Indeed, in many ways the Confederacy devolved into a situation in which, rather than merely competing over the content of policy, the custodians of power became so fragmented that they saw themselves as presiding over separate and self-contained policy areas. In such an environment, pluralism's healthy competition is lost, and various interest groups capture different pieces of the government and shape policies to suit their needs.[25] State's rights governors like Brown and Vance who manipulated the exemption policy as a means of resisting conscription illustrate this technique. As if such challenges were not enough, the Confederate government for one period organizationally fragmented its own power to conscript by granting Colonel John Preston, the erstwhile Superintendent of the Bureau of Conscription authority in only certain states while giving departmental commander General Joseph Johnston control of others.[26]

In addition to this contest between central and state authority, class tensions fragmented power in the conscription effort. The Twenty Slave Law was a controversial exemption measure passed by the Confederate Congress on October 11, 1862, that deferred from military service any planter or overseer with more than twenty slaves. Proponents argued the act was necessary to maintain agricultural production and to prevent slave insurrection. Opponents, including many Southern yeomen, considered the exemption to place an unfair burden for fighting the war on the lower classes. Indeed, some enterprising planters took to dividing their slaves into gangs of twenty, put them on separate tracts of land, and made their sons or other relatives overseers in order to obtain additional exemptions.[27]

The Twenty Slave Law is an example of Paul Escott's assertion that the planters gave primacy to their own rather than national interests. He argues these elites proved "unwilling to make sacrifices or to surrender [their] privileges for independence."[28] Like Governor Brown's insistence on state's rights, the planters' actions are another example of Lowi's contention that "no individual interest group can be expected to take fullest account of the consequences of its own claims."[29]

The Twenty Slave Law generated such a public outcry that it directly contributed to community resistance to conscription. Jaspar Collins, one of the leaders of the anti-government "Free State of Jones," was among those incensed by the legislation. Along with Newt Knight, Collins led a group of between fifty and a hundred like-minded individuals, all linked by family relations and community association and united by dissatisfaction with the status quo, to band together to resist the Confederacy in Jones County, Mississippi. Desertion from the Army and evasion of the draft was tacitly endorsed and supported by this community that became known as "The Free State of Jones." In such areas, anti-government sentiment or general indifference was so powerful "during the last two years of the war as to defeat all efforts to compel slackers to service."[30] The Free State of Jones illustrates that under pluralism, local power, even below the state level, can be leveraged to influence decision-making and challenge central authority.

In response to the rash of criticisms, Congress amended the Twenty Slave Law on May 1, 1863, making it applicable only to overseers on plantations belonging solely to "a minor, a person of unsound mind, a *femme sole* [single woman], or a person absent from home in the military or naval service of the Confederacy." Furthermore, planters were required to swear an affidavit that they had been unable to secure an overseer not liable for military service and to pay $500 for the exemption. To further cut down on abuses, only men who had been overseers prior to April 16, 1862, on plantations that had not been divided since October 11, 1862, could qualify for exemptions. On February 17, 1864, Congress reduced the requirement to fifteen slaves but also required planters with exempted overseers to deliver to the government one hundred pounds of bacon or its equivalent for every slave and to sell surplus produce to the government or to soldiers' families at government prices.[31]

This fifteen slave provision was part of a new Conscription Act that aimed to generate increased military manpower by drastically limiting exemptions and expanding the age limits to between seventy and fifty. It freed conscription officers from the previous limitation of having to accept at face value certificates of physical disability from local physicians who were obviously misrepresenting healthy individuals. It allowed the government to allocate Southern manpower by assigning military details. Under this practice, if the government considered a man to be more valuable in his civil capacity, he could be drafted and then detailed back to his previous job. This new act, however, was poorly implemented, and only 15,820 conscripts joined the Army between January 1 and April

1. About the same number volunteered, but the Bureau of Conscription exempted 26,000 and detailed another 13,000 to critical wartime industries. The result was a net loss of personnel in the military ranks.[32]

Conscription always proceeded along with volunteering, and many men cleverly volunteered as "a method of avoiding service." Colonel Preston complained that the generous furloughs and other incentives offered by recruiting officers made volunteering "a fruitful source of delay in entering and of final escape from service."[33] Other potential draftees nominally met their service obligation in the "bomb proof" state militia units or the various cavalry organizations that "sprang up spontaneously and operated without authority from any source."[34] Even the modest effort of Alabama Governor Thomas Watts in September 1864 to enroll men between the ages of sixteen and fifty-five into a revitalized state militia that would operate as a supersized constabulary under his control, was resisted stridently. Preferring the home-guard system that restricted most militiamen to their home counties, Watts's opponents challenged him to round up deserters instead. Indeed, in a situation similar to Mississippi's Jones County, the northern part of Alabama was teeming with organized bands of deserters that rendered law enforcement there virtually impossible.[35]

In fact the Army suffered its most exasperating losses when such individual soldiers resisted conscription by exercising their power to "vote with their feet." In February 1865, the Bureau of Conscription offered what many thought was a conservative figure in estimating some 100,000 Confederate soldiers had deserted the ranks.[36] The Confederacy could ill-afford this loss of manpower in its struggle against the more heavily populated North. There were several reasons for deserting, and incidences markedly increased as the war progressed.

Some soldiers deserted because of the hardships they personally experienced such as danger, shortages of food and clothing, and irregular or insufficient pay. Some had always been pro-Union or at least lukewarm to the Confederate cause and had been conscripted against their will. When the opportunity presented itself to desert, these soldiers acted on their sentiments.

Increasingly, soldiers deserted not on their own behalf, but out of concern for their loved ones at home. Shortages, marauders, and approaching Federal armies led many Confederate soldiers to desert as a response to letters from home that described the deteriorating home front situation. This phenomenon may in part explain why desertions were noticeably higher among soldiers of the lower classes.[37]

To help mitigate a soldier's inclination to desert, furloughs were often used to ease the difficulty of another hardship or to provide an incentive or reward. For example, the Second Conscription Act required those already in service to remain, but authorized the twelve-month volunteers to receive a 60-day furlough. In spite of this concession, even with an approved furlough, the logistics of getting home was still a problem. As soldiers were stationed further and further from their homes, delays resulting from inadequate transportation could exhaust much of a furlough. For this reason, once soldiers arrived home, they were often reluctant to leave. Some extended their stay without authorization, often accompanying their late return with a concocted story of broken-down transportation or a sickness incurred while at home. Others simply deserted.[38]

Some deserters, especially in the more mountainous areas, would organize themselves into armed bands and collectively resist efforts of the authorities to return them to their units. In many cases, these groups received comfort and support from loved ones in the community. In other cases, deserters preyed upon the vulnerable women and children left behind by soldiers still serving at the front.

Toward the end of the war, desertion became epidemic. In a single day in February 1865, 400 men deserted from Major General Sterling Price's command, and an entire brigade deserted from the Army of Northern Virginia the next month.[39] While most Confederates bravely soldiered on to the war's end, a significant number decided their war was over before the official surrender. The power inherent in an individual's willingness or unwillingness to comply with a policy is certainly an important element of pluralism.

While conscription was no doubt an imperfectly executed system, it did serve to mobilize almost the entire Southern military population. At one time or another, about 750,000 of the approximately 1,000,000 white Southern men between the ages of eighteen and forty-five served in the Confederate Army. About 82,000 of these were drafted. Furthermore, after the Confederate government began the draft, volunteering enjoyed renewed popularity as a means of avoiding the stigma associated with conscription. Facing the much larger population of the North, it is hard to imagine the Confederacy being able to fight like it did if it did not have conscription.[40]

On the other hand, scholars note that while conscription "increased Confederate strength by bolstering the armed forces, ... it also had the compensating deficiency of undermining Confederate morale."[41] President Davis and the Confederate Congress did not have unchecked power in enacting the conscription legislation. In crafting the policy they had to accommodate the power held by other interests at the state and individual levels. Even after this compromise, opponents to conscription used their power to resist the legislation. David Donald overstates the case in asserting that in his effort "to construct a more stable army through conscription, [President Davis] probably lost more than he gained," but the point is well-taken.[42] In the shared power-construct of the pluralist decision-making model, and, even more generally, in a pluralist society, trade-offs abound, and a cohesive policy is difficult to create and implement in such a fragmented environment.

Example 2: The Reflagging of Kuwaiti Tankers in the Midst of the Iran-Iraq War, 1987–1988

The roots of the Iran-Iraq War can be traced to the two sides' religious differences as Shi'ites and Sunnis and their ethnic differences as Persians and Arabs. Iraq's Saddam Hussein was particularly afraid of the threat that the spread of Islamic fundamentalism represented to his Shi'ite population. He also sought to counter Iran's efforts to undermine his regime by seizing key geographic areas that would enhance Iraq's economic and political power. Moreover, Saddam thought Iran had been weakened by the upheaval associated with the 1979 Revolution. Returning from exile, Ayatollah Ruhollah Khomeini had subsequently reduced the size of the armed forces and increased the number of militia units. In the process, many trained officers were replaced by religious leaders with little or no military experience.

On September 22, 1980, Iraq launched a surprise attack on ten Iranian airfields and followed up the strikes with ground attacks into Iran on four separate axes. The air attack yielded disappointing results because many of Iran's best planes were safely sheltered in protective hangars. The ground attacks also made little headway, and by March 1981, the Iraqis had exhausted themselves. Rather than the decisive victory he had hoped for, Saddam had gained just a thin strip of Iranian territory along the border, and, worse, given the Iran time to mobilize its greater population and resources.

The Iran-Iraq War was problematic for the United States and the rest of the international community. Although Saddam's aggression and history of human rights abuses won Iran the moral argument, many feared the spread of Khomeini's influence in the Middle East. Iran received little international support which would cause it significant logistical problems during the war.

In September 1981, Iran seized the initiative and launched a series of attacks. The effort, however, was poorly coordinated and poorly led. In some cases, the Iranians resorted to human wave attacks. Gradually performance improved, and, despite huge losses, the Iranians pushed the Iraqis back. By the end of June 1982, Saddam ordered the withdrawal from most of the Iranian territory his forces had seized earlier. Having repelled the invader, Iran now attempted to defeat Iraq and depose Saddam. In July, the Iranians launched a huge offensive to capture Basra, Iraq's second largest city. This attack failed, but, undeterred, Iran began a new offensive in October with a drive toward Baghdad. Iraq countered this threat with chemical agents.

The war took a new turn in 1984 when Saddam began using his superior air power to attack the shipment of Iranian oil through the Persian Gulf. In February, Iraq attacked the Kharg oil terminal followed by attacks on several tankers with Exocet air-to-surface cruise missiles. Since the Iraqis shipped their oil by pipeline through Turkey, the Iranians could not reciprocate against Iraqi tankers. Instead they attacked the shipping of Iraq's allies, Kuwait and Saudi Arabia. In this phase of the war, Iran maintained the initiative on the ground, while Iraq controlled the skies.

In 1985, Iraq intensified its missile attacks against Iranian cities, including more than forty strikes against Teheran. After Iran was able to acquire its own Scud missiles from Libya, it too began attacking cities, particularly Baghdad. As this "War of the Cities" see-sawed back and forth, Iran's steady improvement in the quality of its commanders allowed it to launch two major offensives simultaneously for the first time in the war. In early 1986 it launched one attack north of Basra and one in the Fao peninsula. This second attack severed Iraq's direct access to the Persian Gulf. An Iraqi attack against Mehran in May was driven back in July with heavy losses, and it appeared that Iran was gaining the upperhand. The United States finally decided that an Iranian victory was against its interests and threatened the stability of the Middle East.[43]

Having reached this conclusion, the US made a move that reflected a "tilt" from the previous effort to remain neutral to a position that clearly supported Iraq. On March 7, the US offered to reflag eleven Kuwaiti vessels and escort under the protection of the US Navy. On March 10, Kuwait accepted the American offer, even though Kuwait had previously approached the Soviet Union with a similar request and received a favorable response. American proponents of the measure saw it as part of an effort to ensure free access to Persian Gulf oil and to deny Soviet influence in the region.[44] The first convoy began on July 21, and by the time this Operation Earnest Will ended in December 1988, the US Navy had escorted 188 reflagged Kuwaiti tankers.[45] The proximate cause to the end of the crisis came on July 3 when the USS *Vincennes* mistakenly shot down an Iranian civilian airliner with 290 passengers on board. In the wake of this disaster, Iran agreed to accept United Nations Security Council Resolution 598 and a cease-fire with Iraq. The Iranian President explained his decision saying the situation has "now gained unprecedented dimensions, bringing other countries into the war and even engulfing innocent civilians."[46]

Numerous agencies made competing arguments as the reflagging policy unfolded and was implemented. The battle between the executive and legislative branches was perhaps the most dramatic, but there were many more. Within the executive branch, the Secretary of Defense and the Department of the Navy held different views. In executing its inspection and documentation functions, the Coast Guard found itself besieged by protests from a variety of labor organizations. In light of this pluralism, the policy and implementation that emerged "would both fail to satisfy our 'friends' in the region, or deter Iranian action."[47]

Secretary of Defense Caspar Weinberger favored reflagging. So did National Security Advisor Frank Carlucci. Interestingly, in his memoirs, Weinberger writes Secretary of State George Shultz "did not share my enthusiasm for this mission," but Shultz recalls in his memoirs that "for once, on an issue involving the use of our military force, Cap Weinberger and I were on the same side—for reflagging."[48] Suffice it to say that President Reagan and his key advisors favored the policy. Instead, the main competition in the executive branch was within the Department of Defense where Weinberger's strong support of reflagging was challenged by the Department of the Navy which would be called upon to execute it. The military's firm commitment to civilian control notwithstanding, clearly the shared power between the makers of policy and the executors of it is a classic example of pluralism.

The Navy feared the reflagging operation would divert a large number of resources from existing, long-term commitments; that it would take funds away from other programs and projects; and that it could produce a loss of lives and ships. Secretary of the Navy James Webb particularly objected to what he felt was the inability to know when "we had won." In an August 7, 1987, memorandum to Weinberger, Webb argued that "there is no definitive action that will be accepted as evidence we have won, or when our commitment will be viewed as having been successfully completed." Webb was also concerned that "by reflagging the Kuwaiti tankers and calling them our own, the American government had not only provoked Iran but overtly titled toward Iraq." He saw no sense that the US had "directly injected itself into the daily workings of a region where violence is the very emblem of its history and where political loyalties shift like powdered sand." For Webb, rather than a well-reasoned strategy, the US was embarking on "a mind-boggling roller-coaster ride."[49]

Indeed, the New York *Times* reported that in spite of the official support of the Joint Chiefs of Staff, "many admirals and generals in and out of Washington admit misgivings about the Administration's reflagging policy because they do not know how long it will take nor where it will lead." The ambivalence of the military is perhaps best captured by General P. X. Kelley, the Marine Commandant and member of the Joint Chiefs, who said, "life is full of lousy options." From those imperfect solutions, Kelley continued, choices have to be made, and in that context, he supported the reflagging. Still, the *Times* article reported, "more than a dozen senior officers spoke of their concerns on the condition they not be identified." Some of these anonymous sources questioned the degree to which the military was involved in the decision-making process, with one admiral saying, "it would be stretching it to say that the Chiefs were in on the decision, or even asked their opinion on it." Others expressed some surprise that Weinberger, traditionally a champion of the conservative use of military force, would support the reflagging. They ascribed this seeming reversal to the political turmoil generated by the Iran-Contra affair. In this case, "politics overcame philosophy," according to one officer.[50] If so, this dynamic would represent pluralism's representation of public opinion as a balance to raw governmental power.

While the strategic policy decision and its operational execution represents one component of the reflagging debate, there were also bureaucratic procedural issues that presented additional opportunities for pluralism. The United States Coast Guard had to approve separate inspection and documentation applications before reflagging could occur, and the initial Kuwaiti interest in reflagging came on December 10, 1986, in the form of an inquiry to the Coast Guard about these requirements. Richard Armitage, Assistant Secretary of Defense for International Security Affairs, testified that because some fifty ships had been reflagged by the Coast Guard over the past four or five years, the service treated the Kuwaiti case "very routinely and it never reached any policy level."[51]

The Coast Guard based its decision-making on preexisting policy and practice. As part of the permissible inspection procedures, the Department of Defense requested a one-year waiver from compliance with those US requirements that exceeded those of international safety conventions. This request was authorized by preexisting national security waivers provided in the law. Under normal procedures, US-flagged vessels must carry a full complement of American officers and a crew of at least seventy-five percent American membership upon leaving a US port. If, however, a vessel is in a foreign port and is deprived of her crew, non-American replacements are allowed until the vessel returns to a US port and can secure American replacements. The Coast Guard considered the Kuwaiti tankers as falling within the bounds of this provision.[52]

Although the Coast Guard considered this a fairly routine matter to be adjudicated largely according to the organizational process model, other actors saw it as a challenge to their special interests.[53] A variety of seamen's groups challenged the Coast Guard interpretation. Frank Drozak, President of the Seafarers International Union, argued that the exemption to the seventy-five percent rule was designed for emergency crew replacement situations and only for the amount of time it would take for a replacement crew to travel to the ship's location.[54] He also objected on the basis that union members were willing and available to serve on the tankers, and were best equipped to operate under conditions related to the national security interests at stake.[55] The representative for the National Maritime Union argued that the reflagging would "adversely affect the American-flag merchant marine and US citizen seamen now and in the future," because waiving the law "makes a farce of our ownership, manning, construction, and safety requirements standards."[56] Captain Robert Lowen, President of the International Organization of Masters, Mates and Pilots, ILA, AFL-CIO, clearly expressed his group's interest in stating that "we are obligated by our own oaths of office to promote and protect our members' jobs and their working and retirement standards."[57] All of these labor-related perspectives were not concerned so much with the reflagging policy per se as they were with how it would be implemented, and labor organizations naturally favored procedures that would bring more business for their members.[58]

Balancing this group was the Reagan Administration's desire for quick action.

Requiring the Kuwaitis to hire American seamen and enter into commercial charter arrangements would slow and complicate the process. The administration did not want to risk Kuwait's rejection of US protection, especially if it resulted in Kuwait turning to the Soviet Union as a more expedient alternative.[59] The Soviets were in fact working on an agreement to reflag or recharter Kuwaiti vessels and on March 2, Kuwait had offered to reflag six tankers under the US flag and five under the Soviet flag.[60]

In cases of such conflicting interests, the pluralist model predicts that the government will act as an umpire, supervising the competition, and compelling a settlement if necessary. In this case, the competition was between the benefits to the US economy and the labor groups of using American crews and ships, and the importance and urgency of protecting national security interests.[61] As Marion Creekmore, Deputy Assistant Secretary of State for Near Eastern and South Asian Affairs explained, "We were dealing here with an issue of high strategic importance. The commercial interest is obviously always important. But in something like this, to try to put it as a condition of our working together with the Kuwaitis for something that we thought advanced major American interests did not seem to be the appropriate approach at the time, nor would I think it was the appropriate approach today."[62] Likewise, Philip Haseltine, Deputy Assistant Secretary of Transportation for Policy and International Affairs testified that "the major issue was not the impact on US maritime policy," but was a decision "made on national security and defense grounds."[63] Michael Rehg argues that because the representatives of these interests—the Secretary of State and the Secretary of Defense—enjoyed closer proximity to the President, they had "more power and influence over the means to reflagging than the organizations opposed to those means."[64] Rehg's conclusion is consistent will the pluralist model's premise that although policy is the product of bargaining between groups, it reflects the interest of the dominant groups.

Another procedural requirement was that the vessels be under ownership by a US citizen or a corporation with fifty-one percent of its directors and operating officers US citizens. To meet this requirement, the Chesapeake Shipping Company of Dover, Delaware was chartered on May 15 and assumed ownership and operation of the eleven reflagged tankers, previously owned by the Kuwait Oil Tanker Company. The Delaware corporation then chartered the tankers back to the Kuwaiti company.[65]

Inspection of the vessels began in May, documentation applications were processed in July, and the escort operations began July 22, 1987. However, the routine flow of these bureaucratic processes was interrupted on May 17 by Iraqi attack on the USS *Stark*. This destroyer was patrolling off the Saudi Arabian coast when it was mistakenly attacked by Iraqi jets. Thirty-seven Americans were killed, and the tragedy brought a new level of public discussion and Congressional opposition to the implementation of the reflagging policy.[66] "The *Stark* was hit and whammo, the white light of public attention fell on that region," explained Senator James Sasser.[67] Indeed, public engagement in a national security issue typically generates increased Congressional involvement in the national security policy process.[68]

The result of this surge in Congressional interest in reflagging unleashed what Secretary Weinberger described as being "in many ways ... the classic battle between the legislative and executive branches of our government."[69] The 100th Congress was a particularly heated iteration of this traditional separation of powers. One contemporary observer wrote that governmental business had "turned into mean-spirited, partisan

The reflagged tanker *Bridgeton* struck a mine on the first escort mission of Operation Earnest Will, raising Congressional and public interest in the operation. Here sailors watch for mines from the USS *Kidd* as the destroyer trails the *Bridgeton* (photograph by PH2 Tolliver; U.S. Navy).

trench warfare between the Democrats and President Reagan backed by his Republican allies." The Iran-Contra Affair exacerbated the situation, leaving the President and Congress fighting to a standstill on many issues. The situation was so heated that Senator Alan Dixon could find "no attempt between Congress and the President to work together." Yet, divisions within Congress gave President Reagan an advantage in this struggle. With a solid 258–176 majority in the House and a less decisive but still promising 54–46 advantage in the Senate, Democrats had hoped to efficiently and productively advance their legislative agenda. The Senate proved particularly vexing to this plan, however, as conservative Democrats often sided with Republicans, creating an even split on many issues. The resulting impasse soon "caused frustrated Democrats to snipe at one another and ... deepened the institutional rivalry between the House and Senate."[70] One way the president's Congressional opponents hoped to gain some leverage was through the provisions of the War Powers Resolution.

The War Powers Resolution was a legislative effort to limit the executive's war-making authority that grew out of Congress's frustration with its ability to influence the prosecution of the Vietnam War. It was passed on November 7, 1973, overriding the veto of President Richard Nixon. The resolution requires the president to immediately report any troop deployments to Congress, and that sixty days later, if Congress has not validated the president's action with a declaration of war, the troops must be withdrawn. If this occurs and the president further refuses or delays to recall the troops, after ninety days Congress has the authority to remove the military forces.[71]

Nixon's veto of the original legislation was indicative of the attitude of subsequent presidents who continued to act unilaterally even after the War Powers Resolution has passed. In this case, the president's advisors were split on the applicability of the act. Even after the *Stark* incident, Weinberger, Shultz, and Carlucci claimed it did not apply since the intention of the reflagging was to deter further military action. On the other hand, Chief of Staff Howard Baker, Attorney General Edwin Meese, and Secretary of the Treasury James Baker recommended compliance. In August, the Senate voted 91–5 to require the Administration to inform Congress of the situation in the Persian Gulf prior to actually reflagging the tankers. President Reagan complied with this "meaningless provision" even before it was passed by the House, but refused to issue a report under the War Powers Resolution. In this case, Reagan argued that escort duty and minesweeping operations in international waters did not involve either hostilities or the possibility of imminent hostilities.[72]

Unwillingly to acquiesce, 110 House Democrats attempted to force compliance with the War Powers Resolution by bringing a suit before a federal court. The court quickly dismissed the suit on December 18, 1987, declaring it was a nonjusticiable political question, and that the plaintiffs' dispute was "primarily with fellow legislators."[73]

In the meantime, Congress's inability to act cohesively gave the President "a strong hand" in this battle between the branches. When the House voted on a ninety-day delay in the reflagging, a similar proposal was filibustered in the Senate. Perhaps even more indicative of the legislative branch's conflict within itself was Senate Resolution 194 which passed by a 54–44 margin on October 27, 1987. This resolution required the president to report within thirty days on his policy. Thirty days after that, under a special rule that precluded a filibuster, the Senate could vote on a joint resolution framing Gulf policy. Thus the resolution guaranteed the matter would come before a vote.[74]

This resolution, however, was riddled with contradictory language. On the one hand, it supported "a continuous US presence in the Persian Gulf," while on the other it expressed "reservations about the convoy and escort operations of the US naval vessels." It also contained an amendment offered by Senator Bob Dole that specified that nothing in the resolution could be construed as limiting the president's constitutional powers as commander-in-chief to use force in self-defense. Another amendment offered by Senator Jesse Helms authorized the Navy to sink any Iranian ship or destroy any Iranian installation that threatened an American warship. Richard Pious concludes the resolution thus "incorporated the most bellicose aspects of the convoy operations and took the most extreme positions on the president's war powers to satisfy the hawks, while at the same time provided for a vote in the future on the entire convoying operation to satisfy the doves."[75]

Eventually President Reagan settled on a policy of limited consultations that pacified his challengers in Congress while sidestepping the War Powers Resolution. "There won't be a fight so long as they keep consulting," House Whip Tony Coelho promised. Although Reagan, like other presidents before him, may have felt he was asserting executive power by ignoring the War Powers Resolution, pluralism suggests such displays come with a cost. In this case, Pious argues that Reagan "alienated members of Congress [and] gave opponents of his policy leverage by allowing them to raise constitutional and legal issues and fault the legitimacy of his actions." The War Powers Resolution may be a dismal failure in the formal means by which it was intended to limit presidential power, but it still indirectly reinforces pluralism's idea of competing and balancing powers.[76] Certainly that is how the policy to reflag and escort Kuwaiti tankers unfolded.

Example 3: The Implementation
Force (IFOR) in Bosnia, 1995

Yugoslavia, always an unnatural composite of various ethnic groups, simply could not sustain itself in the wake of the combined forces of the death of Josip Tito, the end of the Cold War, and the decline of communism. The rise of Slobodan Milošević to power and his agenda of Serb nationalism set "the country on a slow but unrelenting march toward dissolution." Some republics like Slovenia and Macedonia were able to leave the Yugoslav federation relatively peacefully. Croatia's declaration of independence was followed by years of widespread and bloody fighting between ethnic Croats and Serbs, but the bloodiest civil war occurred in multiethnic Bosnia and Herzegovina. Radical nationalists embarked on a horrific and organized campaign of ethnic cleansing. The results were devastating. By 1994, over 200,000 soldiers and civilians were dead or missing. An estimated two million people were either refugees or displaced from their homes. In the summer of 1995, after receiving a healthy infusion of Western training and equipment, the Croatian army drove the Serbian forces out of the Croatian regions of Krajina and western Slavonia. Then a combined Bosniac-Croat offensive began advancing into western and central Bosnia. Heretofore weak Western resolve for military intervention coalesced after a Serb mortar attack devastated a crowded market in Sarajevo, and NATO launched Operation Deliberate Force, an air campaign against the Serbs. These developments brought the warring parties to Dayton, Ohio, for peace talks in November 1995.[77]

Power was so fragmented in the case of Bosnia that the pluralist decision-making model is applicable on many levels. The Bosnian, Croat, and Serb entities all shared power. So did the legitimate national militaries and the paramilitaries. There were elected government officials with tenuous control over charismatic local leaders. The Americans and the Europeans, NATO and the US, the United Nations and NATO, non-governmental organizations and governmental organizations, and many other players all variously competed and cooperated. As an example of the pluralism at work in the international decision-making arena consider Michael Mandelbaum's assertion that "The Americans and the Europeans were each able to veto the policy the other wanted. The United States prevented the implementation of the peace plans; the Europeans blocked all but token bombing. The war dragged on."[78]

As this present study focuses on American decision-making, it will as much as possible confine itself to the pluralist dynamic internal to the United States. In that context, the Bosnia case study is particularly illustrative of a time when the end of the Cold War consensus, an assertive and ideological Republican Congressional majority, and a Democrat president perceived as being weak on foreign policy combined to potentially create a new power relationship in the use of force decision-making dynamic.[79]

Within the US, power was fragmented among a variety of actors. The public was decidedly skittish about the prospects of US involvement in Bosnia. The humanitarian interest was always the most visible concern, yet in a *Time*/CNN poll in October only one-third of Americans indicated believing the US had a moral obligation to stop the fighting in Bosnia.[80] After President Clinton presented his case for the deployment to the

American public in November, an ABC News poll showed slightly over half of Americans now agreed US national interests were at stake in Bosnia. Still, calls to Speaker of the House Newt Gingrich's office ran 976 to 28 against Clinton's policy.[81]

The opinions voiced by Gingrich's constituents translated to Congressional opposition to the Bosnia mission. Already, their gains in the 1994 midterm election cycle had emboldened House Republicans to perceive a public mandate for change. In contrast to the recent growth of peacekeeping operations, to include the ill-fated Somalia operation, the "Contract with America" had promised "no US troops under UN command and restoration of the essential parts of our national security funding to strengthen our own national defense and maintain credibility around the world." Even as the talks in Dayton were about to begin and then as they were in progress, Congress signaled its intentions to act on behalf of what Congressman Dick Armey described as "the voice of the American people ... [that] the nation has gone too far in the direction of globalism."[82]

Two days before the Dayton talks began, the House passed a nonbinding resolution that "there should not be a presumption" that enforcement of a peace agreement "will involve deployment of United States armed forces on the ground in ... Bosnia." The White House pushed back with US mediator Ambassador Richard Holbrooke accusing the resolution's supporters of "doing grave damage to the national interests," being "extremely unhelpful," and serving to "weaken the United States."[83] Then on November 17, with negotiations at Dayton in progress, the House voted largely along party lines to bar President Clinton from sending US ground troops to enforce a Bosnian peace agreement without Congressional approval. The bill was largely symbolic, as it stood little chance of approval in the Senate and certainly there was insufficient Congressional support to override a presidential veto, but State Department spokesman Nicholas Burns still called the House vote a "most unwelcome" development at a "defining moment" in the negotiations. Indeed, a Washington *Post* report called the Congressional action a direct challenge to "the president on the most important foreign policy initiative he has undertaken since the Republicans took control of Congress last January."[84]

Early in the debate, there were the usual rumblings about the War Powers Resolution. On the first legislative day of the new 104th Congress, Senator Bob Dole had introduced a Peace Powers Act that was designed to give Congress a greater say in troop deployments under United Nations auspices. Although it did not result in legislation, it was a clear signal that Republicans desired to limit the president's ability to deploy troops in the era of expanded peacekeeping.[85] When this general concern became specific to the situation in Bosnia, Republican rhetoric intensified.[86] Senator Richard Luger, for example, declared, "As a practical matter, this is going to require a Congressional vote. It won't happen without a vote."[87] Sensing these pressures from Congress, President Clinton mounted a concerted effort to safeguard what he considered his own constitutional powers.[88]

Recognizing the potential of strong Republican opposition, but at the same time being unwilling to fully acknowledge the War Powers Resolution, the Clinton Administration made repeated use of the language of consultation. The president as well as top officials such as Vice President Al Gore, Chief of Staff Leon Panetta, Secretary of Defense William Perry, and Secretary of State Warren Christopher all indicated that Congress would be consulted after an agreement was reached at Dayton. In reality, President Clinton only paid lip service to any substantive consultation. Instead, using what Ryan Hendrickson calls "a novel strategy ... of tacit consultation with Congress," Clinton seized the initiative.[89]

Opponents of the deployment quickly found themselves on their heels as the Dayton peace talks progressed. In a November 13 letter, President Clinton stated his case, telling Gingrich "Congressional support for US participation is important and desirable, although, as has been the case with prior presidents, I must reserve my constitutional prerogatives in this area." Clinton proceeded accordingly and created such momentum that he made the deployment all but inevitable. In fact, even before he requested Congressional support (rather than permission), Clinton made it clear there would be "a requirement for some early pre-positioning of a small amount of communications and support personnel."[90] Once Clinton had the Dayton Peace Agreement in hand, Congressman Floyd Spence lamented, "The proverbial train has left the station and our troops are already on board." Senator Dole reluctantly admitted, "Congress cannot stop this troop deployment," and Senator John McCain added he "was opposed" to the Bosnian policy, "but the president made a commitment and if we reversed it, it would be a tremendous blow to the credibility of the United States."[91]

In this regard, Bosnia is another example of the fact that Congress, when "faced with the unenviable position of denying a president freedom of action in time of crisis, has generally given the president the authority he sought."[92] Many observers believe Congress knows exactly what it is doing through this deference. "Congress's preferred tactic when confronted with knotty foreign policy problems," claimed a contemporary article in *Time*, "is to make lots of noise but leave the hard calls to the President." In pursuing this course, Democratic Congressman Lee Hamilton accused his Republican colleagues of ignobly posturing themselves to have it both ways. "If [President Clinton] succeeds, they'll praise him," Hamilton said. "If he fails, they'll criticize him." A Republican congressional staff member more sinisterly concurred, musing, "If the Commander in Chief wants to hang himself, who are we to take away his rope?"[93]

Instead, both the House and Senate, in spite of the serious reservations of many of their members over the Bosnia policy, choose not to test the President on constitutional war powers issues. Instead, in a deft political maneuver, the House voted 287–141 and the Senate 69–30 in favor of measures that expressed some opposition to President Clinton's policy but supported the troops. Calculating their actions from "a political public relations standpoint, both chambers sought to wash their hands of the policy and remain patriotic at the same time." It was a politically safe course that allowed lawmakers to justify their votes to the voters back home. [94]

If President Clinton was able to muscle his way past Congressional opposition to deploy the troops, Congress still controlled the budget. With an estimated price tag of first $1.5 billion and then $2 billion, Congress had leverage which it was quick to flex. When President Clinton threatened to veto a military appropriations bill that provided for $7 billion more than he requested, Republican Congressional leaders "bluntly told him that if he did, he'd have trouble getting money for Bosnia."[95] Such tough talk faded, however, as the deployment became a reality. Senator Dole reminded his colleagues, "It is time for a reality check in Congress. If we would try to cut off funds, we would harm the men and women in the military who have already begun to arrive in Bosnia." Congressional blustering aside, as William Banks and Jeffrey Straussman note, "When troops are on the ground, the sword drives the purse, not the other way around."[96] Hendrickson also notes that within the context of their important policy objective of obtaining a balanced budget, Republicans could ill-afford to appear obstructionist in the public eye if a government shut-down resulted from the budgetary implications of the Bosnia debate.[97] Although Congressional control over the budget is one of its most powerful tools, it is "a blunt instrument and a difficult one to wield when US armed forces are already in combat."[98] Consistent with the pluralist model, Congress's budgetary power was balanced by the powers of other actors.

Indeed, pluralism predicts that policy will reflect the interest of the dominant group, and the experience with Bosnia suggests that while the president still must contend with the strong constitutional role accorded Congress, he retains the upperhand in decision-making about the use of military force. Much of this advantage is because of the institutional weaknesses such as diffuse power bases, divergent member preferences, and slow decision-making processes that inhibit decisive Congressional action. This condition is particularly prevalent when attempting to set strategic policy. Congress has been more successful in setting structural policy, but writ large, the president holds a historical advantage in controlling decisions about deploying US forces abroad.[99]

Although President Clinton was able to use this advantage to push his policy through, the fragmented power under which the US intervention in Bosnia was brokered resulted in several weaknesses in "IFOR," the multi-national NATO-led military Implementation Force for the peace agreement. The first was the fear of mission-creep that proscribed a very narrow mandate for IFOR. The second was the reluctance to enter an open-ended military commitment that led to the one-year mission. Both these factors would impact implementation of the peace agreement and show how the power-sharing incumbent in the pluralist decision-making model can lead to satisficing.

The Implementation Force was authorized as a peace enforcement operation under Chapter VII of the United Nations Charter by UN Security Council Resolution 1031 on December 15, 1995. As outlined by the Dayton Peace Agreement, IFOR's military mandate was a narrow one. Its primary tasks were to establish a durable cessation of hostilities, ensure force protection, and establish lasting security and arms control measures.[100] Pursuant to these ends, IFOR had responsibility for such tasks as enforcing the zone of separation between the former warring parties and monitoring the withdrawal of heavy weapons into designated cantonment areas. It was also assigned supporting tasks to be done "within the limits of its assigned principal tasks." These supporting tasks included: "to help create secure conditions for the conduct by others of other tasks," "to assist the UNHCR [United Nations High Commissioner for Refugees] and other international organizations in their humanitarian missions," and "to observe and prevent interference with the movement of civilian populations, refugees, and displaced persons."[101]

The mandate was written in a way that allowed IFOR "to go about its business insulated from the more imprecise and therefore potentially troublesome elements of the [Dayton Peace Agreement]."[102] Such thorny tasks as negotiating and implementing military regional-stabilization measures, military confidence- and security- building measures, and the conduct of verification inspections were assigned to the Organization for Security and Cooperation in Europe. Also conspicuously absent from IFOR's mandate were any specified law enforcement or police responsibilities. Instead, the UN International Police Task Force (IPTF) was responsible for law enforcement, and the UN Security Council authorized a 1,721-member civilian police (CIVPOL) operation in December

A Bradley Fighting Vehicle leads a column of IFOR vehicles across a US Army pontoon bridge over the Sava River from Zupania, Croatia, into Bosnia on December 31, 1995 (photograph by Senior Airman Edward Littlejohn; Department of Defense).

1995. Even the IPTF's mandate was limited to monitoring, advising, and training Bosnian police. It had no executive authority to investigate, arrest, or perform other police functions.[103]

As a result of the sharp civil-military distinctions in the Dayton Peace Agreement and IFOR's unwillingness to move beyond its limited mandate, virtually no individuals were arrested for possible war crimes in the first two years after the end of the fighting. International oversight of policing fell entirely upon the IPTF, but the IPTF had little authority and took months to get in place. As a result, law and order were largely left to the suspect ethnic police forces who all resisted international efforts to track down and arrest persons indicted as war criminals by the International Criminal Tribunal for the Former Yugoslavia. The situation was such that Serb President Slobodan Milošević felt confident enough to caution Supreme Allied Commander, Europe General Wesley Clark, "I warned you last time not to go after Serb 'war criminals.' Yet you still talk about this. It is dangerous."[104] It was not until July 1997 that NATO troops seized their first indicted war criminals, and, even then, principal indicted figures such as Bosnian Serb leaders Radovan Karadžić and Ratko Mladić, as well as the once smug Milošević, eluded being brought to justice for what most considered an unacceptably long time.

Such limited mandates as IFOR's are consistent with the goal of what Donald Snow calls "conflict suppression." Military forces involved in this activity seek to ensure fighting does not resume. It is a task the military understands and is good at, and it can usually be performed with minimum casualties. In the case of Bosnia, such an approach helped mitigate military, public, and Congressional apprehensions about a potentially dangerous mission. On the other hand, conflict suppression as a sole strategy works so long as the military force is in place, but does little or nothing to alter the underlying problems that were the original cause of the violence. This omission makes it very difficult to determine an end point at which the mission can be declared accomplished.[105] In the case of IFOR, policy-makers tried to work around this limitation by opting for a time-based end state rather than a much more effective conditions-based one.

Secretary of Defense William Perry spoke to Congress on November 30, 1995, to address this situation and insisted that the Bosnia operation had a clear exit strategy. He said,

> We believe the mission can be accomplished in one year, so we have built our plan based on that timeline. This schedule is realistic because the specific military tasks in the agreement can be completed in the first six months, and thereafter IFOR's role will be to maintain the climate of stability that will permit civil work to go forward. We expect that these civil functions will be successfully initiated in one year. But, even if some of them are not, we must not be drawn into a posture of indefinite garrison.[106]

Amos Jordan and his colleagues describe this exit strategy in decidedly pluralist terms: "believing that the public would be unwilling to support a more substantial commitment, Congress and the president agreed to a one-year deployment."[107] The result according to Michael Mandelbaum was that "the chief purpose of an American expeditionary force in Bosnia would be to leave as soon as possible, with as few casualties as possible, rather than to do whatever was necessary, for as long as necessary, to keep (or make) peace."[108] This artificial time-based objective may have helped allay Congressional concerns of an open-ended mission, but it also had detrimental operational impacts.

One problem was that in an effort to comply with the one-year time limit, the Clinton Administration pressed successfully for early and frequent elections at each level of governance. Many observers warned that holding elections so soon after the cessation of hostilities would merely consolidate the power of those nationalist extremists dedicated to resisting reconciliation among Bosnia's ethnic communities. In most cases, that is exactly what happened. Each ethnic group's most belligerent and nationalistic political party — the Muslim Party of Democratic Action, the Croatian Democratic Union, and the Serbian Democratic Party — swept the legislative elections at both the national and entity level. Ambassador Holbrooke, who had labored so hard to make the Dayton Peace Agreement a reality, lamented, "The election strengthened the very separatists that had started the war."[109]

A second problem was that implementation of the civilian aspects of the Dayton Peace Agreement lagged behind progress on the military side.[110] The Implementation Force assumed formal transfer of authority on December 20, 1995, and by mid–January 1996 had cleared the four-kilometer zone of separation between the former belligerents. Between February and mid–March, IFOR supervised the redeployment of forces on either side of the "inter-entity boundary line." Under IFOR supervision, the former warring parties proceeded to place their heavy weapons in designated cantonment sites and demobilize some 300,000 combatants. Even though most of the formal military tasks were complete, the security situation in Bosnia was clearly not self-sustaining, and a departure of the NATO troops would almost certainly have led to a resumption of conflict between the former warring parties.[111]

As a result, when IFOR's mandate expired in December 1996, NATO did not withdraw from Bosnia. Indeed at the time, few forthright observers had any hopes for mean-ingful and enduring progress being achieved in one year, and commentators said "it came as no surprise" when the mission was extended.[112] The Implementation Force was suc-ceeded by another NATO-led force, the Stabilization Force (SFOR), which was given an 18-month mission. Yet as IFOR drew to a close, thoughtful commentators noted that "The United States should base its decision on continued participation in ... [Bosnia] on whether US national objectives and interests have been achieved, not on a rigid timetable."[113] While such an argument represented sound strategic thought, it also reopened fears of mission creep. Supreme Allied Commander, Europe General George Joulwan pointedly asked NATO if SFOR would be tasked to pursue war criminals, escort returnees, or engage in police work. The North Atlantic Council (NAC) unequivocally told Joulwan "no." Nonetheless, within eight months, by July 1997, SFOR was engaged increasingly in var-ious aspects of all three of these activities. [114] This mission creep trend, as well as the peacekeeping presence in Bosnia, would continue. The Stabilization Force was extended in June 1998, and this time, no end date was specified. It was not until 2004, almost ten years after the original "one year mission," that NATO decided it had accomplished its objective in Bosnia and declared "SFOR has been brought to a successful end."[115] That SFOR was replaced by a new international presence of 7,000 personnel supported by NATO — the European Force or EUFOR — lends some ambiguity to this assertion.

Conditions-based endstates and the broad mandates required for long-term state-building rather than mere conflict suppression have gained grudging acceptance, as the subsequent US experience in Kosovo would suggest.[116] Pluralist decision-making wars against such outcomes, however, because no party wields sufficient power to decisively dictate policy. The result in Bosnia was an initial peacekeeping force that included built-in shortcomings that had to be corrected over time as proponents of the US involvement consolidated power.

Utility of the Pluralist Model

In 1863, Governor Joseph Brown was railing against what he saw as the "imperial" presidency of Jefferson Davis. Using the same description, in 1973, Arthur Schlesinger wrote *The Imperial Presidency* in which he expressed his fears that the modern office had grown beyond its constitutional limits. The War Powers Resolution of that same year was an effort to abate this perceived abuse of presidential power. Such fears notwithstand-ing, in 1998, Michael Smith argued that " in the past twenty years, we have moved from the 'imperial presidency' to the 'imperial Congress,' and finally to a rough stalemate between the two institutions today, a situation in which each side depends on the other when committing the United States to a major military operation."[117] Pluralism shares this view of equilibrium among groups as the natural state of affairs.[118]

Yet such power-sharing can make decision-making and policy implementation difficult and suboptimal. In critical national security situations involving the use of force, this limitation can be dangerous. It is even more complicated when one remembers that the president and Congress are just two of many groups that share power in the plural-ist model. To this end, David Davenport described a "new diplomacy" in which

> not only have the actors changed, but so also have the methods. Technology has enabled the cre-ation of new and rapid means of communication. The emergence of CNN and other worldwide

media has created instant marketing for global agendas. Globalization of the world economy has given rise to more integrated approaches to international concerns. And the spread of democracy around the world has created a greater sense of expectation, even entitlement, in policy-making of all kinds. The mantle of international leadership is no longer conferred by economic and military power alone; instead, the power of ideas, and how they are communicated and marketed, has come to the fore.[119]

The actors and methods have only increased since the time of Davenport's observation in 2002.

In such a decentralized environment, decision-making can be frustrated if the various actors pursue only their power and interests, and disregard their roles and responsibilities to the greater good. This requires that within the competition of pluralism, there also must be some division of labor that generates unity of effort. Concerning the president and Congress, for example, Smith considers the ideal situation to be for the president to provide the sense of purpose and urgency in ordering forces overseas, to fulfill his role as commander-in-chief in the chain of command during the deployment, and to represent US interests internationally. For its part, Congress provides the venue for the intensive deliberation and debate that must accompany questions of such importance and provides the political support to US forces in harm's way. But in the periods of divided government that appear to increasingly be the norm in American government, Smith notes "this division of labor can become a major point of contention." Without the unifying Cold War-type external threat to induce members of Congress and the president to play these cooperative roles, Smith argues that "they will have to rely on more elusive catalysts: character and leadership."[120] Such a prescription is certainly consistent with pluralism's assumption that power resources can be substituted for one another, but generating such leadership has proven easier said than done.

In its most efficient form, pluralism provides checks and balances that prevent abuses and excesses of power and results in policy that is "a reasonable approximation of society's preferences."[121] This statement implies a certain cohesiveness in society that is capable of producing a discernible preference. If the society cannot agree on what this collective preference is, the result may be a certain degree of decision-making paralysis which is always a danger with pluralism.

Epilogue

It is hard to imagine a more difficult decision for any leader than to commit the people under his care into harm's way. Thus the decision to use military force would seem to be one made only after the most deliberate and exhaustive process to determine the best course of action. This seemingly obvious statement proves problematic on at least two accounts. First, for a variety of reasons, the process itself is subject to many variables that war against the rational, logical approach that the situation seems to require. A lack of time, human frailty, imperfect information, stress, and mistakes all degrade the decision-making process. Second, the qualifier "best" is subjective and means different things to different people. Obviously when more than one actor is involved in the process, the different interpretations of "best" must somehow be reconciled, negotiated, or bullied into acceptance, but even in the cases of a single decision-maker, what he considers "best" reflects his own perspective and may not represent more than personal opinion. Decision-making is thus as much art as it is science, and therefore a fitting subject for analysis and study.

Indeed, much of decision-making as an academic discipline is the result of trying to bring order and explanation after the fact to what was indiscernible as such as the actual event unfolded. It is unlikely that a decision-maker posed with a situation enters the task saying to himself, "I think I shall use the poliheuristic approach to this particular problem." Even those decision-making models such as the small group that are the result of a particular leader's conscious choice are subject to unforeseen subsequent developments such as the phenomenon of groupthink. Thus, the models discussed in this and other works may be of more utility to understand than to make a decision.

This is not to say that the decision-maker does not also benefit from a familiarity with the various models and their strengths and weaknesses. In this way forewarned is forearmed, and the decision-maker enters with eyes wide open into a process that is imperfect at best. If we accept the seemingly intuitive conclusion that the rational actor model offers the best chance of an informed and optimal choice, and we accept the equally compelling argument that much, if not most, decision-making deviates from this process, than it would seem that more knowledge of those deviations would be helpful. For example, a decision-maker with an appreciation of the prospect theory would be aware of the impact his domain has on his perception of the situation and not be as unwittingly susceptible to its influences. Even with this awareness, he may still make the same decision, but at least he would do so consciously and deliberately. More knowledge means more control of the variables, and more control of the variables means better decisions. When the stakes are as high as they are with decisions involving the use of force, that is an important advantage.

Chapter Notes

PREÏFACE

1. Bill Clinton, "Why Bosnia Matters to America," *Newsweek*, November 13, 1995, 55.

2. Nikolas Gvosdev, "The Realist Prism: Obama's 'Just Enough' Doctrine in Libya," *World Politics Review*, April 15, 2011, *http://www.worldpoliticsreview.com/articles/8535-/the-realist-prism-obamas-just-enough-doctrine-in-libya* (accessed March 9, 2013).

INTRODUCTION

1. Barton Bernstein, "Secrets and Threats: Atomic Diplomacy and Soviet-American Antagonism," in *Major Problems in American Foreign Relations, Volume II: Since 1914*, ed. Thomas Paterson and Dennis Merrill (Lexington, MA: D. C. Heath, 1995), 268. It should be noted that Bernstein does not specifically mention poliheurism in his analysis.

2. See Rose McDermott, *Risk-taking in International Politics: Prospect Theory in American Foreign Policy* (Ann Arbor: University of Michigan Press, 2001), Chapter 3.

3. Graham Allison, "Conceptual Models and the Cuban Missile Crisis," *The American Political Science Review* 63, No. 3 (September 1969): 689–718.

CHAPTER 1

1. Joshua Goldstein and Jon Pevehouse, *International Relations* (New York: Longman, 2010), 105.

2. Graham Allison, *Essence of Decision: Explaining the Cuban Missile Crisis* (Boston: Little, Brown, 1971), 30.

3. Alex Mintz and Karl DeRouen, *Understanding Foreign Policy Decision Making* (Cambridge: Cambridge University Press, 2010), 58.

4. Mintz and DeRouen, 58; Rose McDermott, *Risk-Taking in International Politics: Prospect Theory in American Foreign Policy* (Ann Arbor: University of Michigan Press, 2001), 117; Russell Bova, *How the World Works: A Brief Survey of International Relations* (New York: Longman, 2010), 89.

5. Greg Cashman, *What Causes War? An Introduction to Theories of International Conflict* (New York: Lexington Books, 1993), 77–78.

6. For a challenge to this popular image of Truman see D. Clayton James, "Harry Truman: The Two War Chief," in *Commanders in Chief: Presidential Leaders in Modern Wars*, ed. Joseph Dawson (Lawrence: University Press of Kansas, 1993), 107–126.

7. David McCullough, *Truman* (New York: Simon & Schuster, 1992), 437.

8. James, 110.

9. Dwight Eisenhower, *Crusade in Europe* (Garden City, NY: Doubleday, 1948), 441–442. See also McCullough, 428 and 437.

10. McCullough, 437.

11. Harry Truman, "Address Before a Joint Session of the Congress, April 16, 1945," *http://www.trumanlibrary.org/ww2/stofunio.htm* (accessed October 8, 2012).

12. Mintz and DeRouen, 57.

13. *Public Papers of the Presidents, Harry S. Truman, 1945*, 212.

14. Harry Truman, *Memoirs by Harry S. Truman*, vol. 1 (Garden City, NY: Doubleday, 1955), 10.

15. Robert Strong, *Decisions and Dilemmas: Case Studies in Presidential Foreign Policy Making Since 1945* (Armonk, NY: M. E. Sharpe, 2005), 1–2.

16. Blaine Browne, "MANHATTAN Project," in *World War II in Europe: An Encyclopedia*, ed. David Zabecki (New York: Garland, 1999), 112; Susanne Teepe Gaskins, "Groves, Leslie," in *World War II in Europe: An Encyclopedia*, ed. David Zabecki (New York: Garland, 1999), 327–328.

17. Ibid.

18. Barton Bernstein, "Secrets and Threats: Atomic Diplomacy and Soviet-American Antagonism," in *Major Problems in American Foreign Relations, Volume II: Since 1914*, ed. Thomas Paterson and Dennis Merrill (Lexington, MA: D. C. Heath, 1995), 267.

19. Strong, 6.

20. Truman, 415.

21. Roy Appleman, et al., *Okinawa: The Last Battle* (Washington, DC: Center of Military History, 1948), 473–474.

22. Arthur Cyr, "Atomic Bomb, Decision to Use Against Japan," in *World War II in the Pacific: An Encyclopedia*, ed. Stanley Sandler (New York: Garland, 2001), 93.

23. Truman, 419. See also Strong, 14.

24. Mintz and DeRouen, 58.

25. Truman, 87.

26. Cashman, 77.

27. Quoted in McCullough, 394.

28. Truman, 417.

29. McCullough, 437.

30. Strong, 14.

31. Bernstein, 266.

32. *Public Papers of the Presidents, Harry S. Truman, 1945*, 212.

33. Quoted in McCullough, 395.

34. Truman, 419 and Strong, 17.

35. "Notes of Meeting of the Interim Committee, June 1, 1945. Miscellaneous Historical Documents Collection," The Harry S Truman Library and Museum, 8–9. http://www.trumanlibrary.org/whistlestop/study_collections/bomb/large/documents/index.php?documentdate=1945–06–01&documentid=40&studycollectionid=abomb&pagenumber=8 (accessed December 12, 2012).

36. Bova, 89.

37. Strong, 7 and "Memorandum by Ralph A. Bard, Undersecretary of the Navy, to Secretary of War Stimson, June 27, 1945," U.S. National Archives, Record Group 77, Records of the Chief of Engineers, Manhattan Engineer District, Harrison-Bundy File, folder #77, "Interim Committee, International Control."

38. Strong, 6.

39. Ibid., 17.

40. Truman, 420.

41. Cyr, 94.

42. James, 109–110.

43. Tsuyoshi Hasegawa, *Racing the Enemy: Stalin, Truman, and the Surrender of Japan* (Cambridge, MA: Belknap Press, 2005), 297. See also ibid., 186.

44. *Public Papers of the Presidents, Harry S. Truman, 1945*, 212.

45. Ibid.

46. Truman, 426.

47. Hasegawa, 297.

48. Bova, 89.

49. H. Richard Yarger and Georghe Barber, "The US Army War College Methodology for Determining Interests and Levels of Intensity," (Carlisle Barracks, PA: US Army War College, 1997), 8.

50. Yarger, 8.

51. David Rapoport, "The Four Waves of Modern Terrorism," *http://www.international.ucla.edu/media/files/Rapoport-Four-Waves-of-Modern-Terrorism.pdf* (accessed March 11, 2013), 56–61. Rapoport notes that "cold war concerns sometimes led the United States to ignore its stated distaste for terror" in places like Nicaragua and Angola. Rapaport, 58.

52. Bruce Jentleson, "The Reagan Administration and Coercive Diplomacy: Restraining More Than Remaking," *Political Science Quarterly* 106, no. 1 (Spring 1991): 63.

53. Rapoport, 58.

54. Jentleson, 63.

55. Cynthia Combs, *Terrorism in the Twenty-First Century* (Boston: Pearson, 2013), 112.

56. Weinberger, 187.

57. "Joint News Conference by Secretary Shultz and Secretary Weinberger of April 14, 1986," *Department of State Bulletin* (June 1986), 5.

58. Jentleson, 58.

59. Caspar Weinberger, *Fighting for Peace: Seven Critical Years in the Pentagon* (New York: Warner Books, 1990), 189.

60. Amos Jordan, et al., *American National Security* (Baltimore: Johns Hopkins University Press, 2009), 276. See also Robert Art, "The Four Functions of Force," in *International Politics: Enduring Concepts and Contemporary Issues*, ed. Robert Art and Robert Jervis (New York: Pearson, 2009), 133.

61. Alexander George, et al., *The Limits of Coercive Diplomacy* (Boston: Little, Brown, 1971), 18.

62. Barry Blechman and Stephen Kaplan, *Force Without War: US Armed Forces as a Political Instrument* (Washington, DC: Brookings Institution, 1978), 13.

63. George, 18.

64. Bova, 89.

65. FM 5–0, *Army Planning and Orders Production* (Washington, DC: Department of the Army, 2005), 3–1.

66. FM 6–0, *Mission Command: Command and Control of Army Forces* (Washington, DC: Headquarters, Department of the Army, 2003), 2–4.

67. Joseph Stanik, *El Dorado Canyon: Reagan's Undeclared War with Qaddafi* (Annapolis: Naval Institute Press, 2002), 152 and Bolger 411.

68. FM 5–0, 3–53–3-55.

69. Stanik, 153; Bolger, 388, 411, and 414.

70. W. Hays Parks, "Crossing the Line," *Proceedings* (November 1986), 45.

71. Weinberger, 189.

72. Ibid., 198.

73. Bolger, 429.

74. Weinberger, 197.

75. Ibid., 199.

76. Jentleson, 63.

77. Brian Jenkins, "Defense Against Terrorism" *Political Science Quarterly* 101, no. 5 (1986): 786.

78. Jentleson, 64.

79. Resolution 660. http://daccess-dds-ny.un.org/doc/RESOLUTION/GEN/NR0/575/10/IMG/NR057510.pdf?OpenElement (accessed December 1, 2012).

80. Robert Scales, *Certain Victory: The US Army in the Gulf War* (Washington, DC: Office of the Chief of Staff, United States Army, 1993), 50–51.

81. Resolution 678. http://daccess-dds-ny.un.org/doc/RESOLUTION/GEN/NR0/575/28/IMG/NR057528.pdf?OpenElement (accessed December 1, 2012).

82. Doughty, 717–720.

83. Ibid., 722–725.

84. Ibid., 725.

85. Norman Schwarzkopf, *It Doesn't Take a Hero* (New York: Bantam, 1992), 497.

86. Dick Cheney, *In My Time: A Personal and Political Memoir* (New York: Threshold Editions, 2012), 225.

87. George H. W. Bush, *All the Best, George Bush: My Life in Letters and Other Writings* (New York: Scribner, 2013), 514.

88. Bova, 89.

89. Colin Powell, *My American Journey* (New York: Random House, 1995), 512.

90. Bova, 89.

91. Cheney, 226.

92. Schwarzkopf, 497.

93. Bush, 514.

94. Bova, 90.

95. Mintz and DeRouen, 57–58.

96. Ibid., 7.

97. Quoted in McDermott, 185.

98. Cashman, 78–79.

CHAPTER 2

1. See Daniel Kahnemsn, and Amos Tversky, "Prospect Theory: An Analysis of Decision under Risk," *Econometrica* 47, no. 2 (March 1979): 263–292.

2. Rose McDermott, *Risk-Taking in International Politics: Prospect Theory in American Foreign Policy* (Ann Arbor: University of Michigan Press, 2001), 4. Hereafter McDermott, *Risk-Taking*.

3. Ibid., 37.

4. Idid., 40.

5. Ibid., 42.

6. Kahneman and Tversky, 279.

7. McDermott, *Risk-Taking*, 35 and Rose McDermott, "Prospect Theory in Political Science: Gains and Losses

from the First Decade," *Political Psychology* 25, no. 2 (April 2004): 290. Hereafter McDermott, "Prospect Theory."

8. McDermott, *Risk-Taking*, 175.

9. Ibid., 170.

10. Ibid., 18.

11. See, for example, Russell Weigley, *History of the United States Army* (New York: Macmillan, 1967), 68.

12. Douglas Southall Freeman, *George Washington: A Biography*, vol. II (New York: Charles Scribner's Sons, 1948), 376.

13. March 3, 1777, John Fitzpatrick, ed., *The Writings of George Washington*, 39 vols. (Washington, DC: Government Printing Office, 1931–1944), VII, 223–234.

14. Robert Doughty, et al., *American Military History and the Evolution of Western Warfare* (Lexington, MA: D. C. Heath, 1996), 41.

15. Maurice Matloff, *American Military History* (Washington, DC: Office of the Chief of Military History, 1973), 66.

16. George Washington to John A. Washington, December 18, 1776, The George Washington Papers at the Library of Congress, 1741–1799, http://memory.loc.gov-/cgi-bin/query/r?ammem/mgw:@field(DOCID+@lit(gw060307)).

17. Thomas Paine, *The American Crisis* (London: James Watson, 1835), 3.

18. Ibid., 5.

19. Ibid., 4.

20. David Fisher, *Washington's Crossing* (Oxford: Oxford University Press, 2006), 142.

21. *Journals of the Continental Congress*, Volume 6 (Washington, DC: U.S. Government Printing Office, 1906), 1032.

22. Wolcott to Adams, January 1, 1777, Papers of John Adams, vol. 5, Document number: PJA05d034, http://www.masshist.org/publications/apde/portia.php?id=PJA05d034, Massachusetts Historical Society.

23. Douglas Southall Freeman, *George Washington: A Biography*, vol. IV (New York: Charles Scribner's Sons, 1951), 336.

24. Reed to Washington, December 22, 1776, in *The Papers of George Washington*, Revolutionary War Series 7, ed. Philander Chase (Charlottesville: University Press of Virginia, 1997), 415.

25. Richard Ketchum, *The Winter Soldiers: The Battles for Trenton and Princeton* (New York: Henry Holt, 1999), 176.

26. Doughty, 43.

27. Ibid., 42.

28. Washington to Reed, December 23, 1776, Chase, 423.

29. Matloff, 67.

30. Ibid., 68.

31. Freeman, vol. IV, 322–324.

32. Ibid., 331.

33. Matloff, 69–70.

34. David Bonk, *Trenton and Princeton 1776–77: Washington Crosses the Delaware* (Oxford: Osprey, 2009), 48.

35. Henry Knox to Mrs. Knox, January 7, 1777, quoted in Freeman, vol. IV, 358.

36. Matloff, 69.

37. Allan Millett and Peter Maslowski, *For the Common Defense: A Military History of the United States of America* (New York: Free Press, 1984), 66.

38. David Skaggs, "The Generalship of George Washington," *Military Review* (July 1974), 8.

39. Abraham Lincoln to Joseph Hooker, telegram, June 10, 1863, CWAL, Volume VI, p. 257.

40. George Meade, *The Life and Letters of George Gordon Meade Major-General United States Army*, vol. 2 (New York: Charles Scribner's Sons, 1913), 3.

41. Ibid., 4

42. Joseph Persico, *My Enemy, My Brother: Men and Days of Gettysburg* (New York: Da Capo Press, 1996), 73.

43. Meade, vol. 2, 11.

44. Ibid., 4.

45. Charles Benjamin, "Hooker's Appointment and Removal," in *Battles and leaders of the Civil War*, vol. III (New York: The Century Company, 1888), 243; Edward Stackpole, *They Met at Gettysburg* (Mechanicsville, PA: Stackpole Books, 1982), 82–84.

46. Kahneman and Tversky, 274.

47. Meade, vol. 2, 8.

48. Stackpole, 85.

49. Meade, vol. 2, 12.

50. Richard Sauers, *Meade: Victor of Gettysburg* (Washington, DC: Brassey's, 2003), 57.

51. Thomas Nelson Page quoted in J. G. Randall, *The Civil War and Reconstruction* (Boston: D. C. Heath, 1937), 523–524.

52. Freeman Cleaves, *Meade of Gettysburg* (Norman: University of Oklahoma Press, 1960), 172–173.

53. Bruce Catton, *Glory Road* (New York: Doubleday, 1954), 327–328.

54. Cleaves, 169.

55. See 1 Corinthians 3:15 NIV.

56. Quoted in Jacob Hoke, *The Great Invasion* (Dayton: W. J. Shuey, 1887), 440–441.

57. Quoted in Cleaves, 172.

58. Hoke, 442.

59. Catton, 327–328.

60. Hoke, 442–443.

61. Catton, 327–328.

62. Cleaves, 171.

63. Meade, vol. 2, 308.

64. Michael Burlingame, *Abraham Lincoln: A Life*, Volume II (Baltimore: Johns Hopkins University Press, 2008), 512.

65. Lincoln to Meade, July 14, 1863, in *Collected Works of Abraham Lincoln*, ed. Roy Basler, 9 vols. (New Brunswick, NJ: Rutgers University Press, 1955), 6: 327–328. Hereafter *CWAL*.

66. Stackpole, 306–307.

67. General Orders No. 68, July 4, 1863, *War of the Rebellion: Official Records of the Union and Confederate Armies*, 128 volumes (Washington, DC: Government Printing Office, 1880–1901), 27 (3): 519. Hereafter *OR*.

68. Lincoln to Halleck, July 6, 1863, *OR* 27 (3): 567.

69. Meade, vol. 2, 307.

70. Andrew Kydd, "Methodological Individualism and Rational Choice," in *The Oxford Handbook of International Relations*, ed. Christian Reus-Smit and Duncan Snidal (New York: Oxford University Press, 2010), 436.

71. Michael Burlingame and John R. Turner Ettlinger, eds., *Inside Lincoln's White House: The Complete Civil War Diary of John Hay* (Carbondale: Southern Illinois University Press, 1999), 64–65 (July 19, 1863).

72. Cleaves, 178.

73. Meade, vol. 2, 309.

74. Ibid., 309.

75. Burlingame and Ettlinger, eds., 63 (July 15, 1863).

76. *CWAL*, Volume VI, p. 341 (Letter from Abraham Lincoln to Oliver O. Howard, July 21, 1863).

77. Carl von Clausewitz, *On War*, ed. Michael Howard and Peter Paret (Princeton: Princeton University Press, 1984), 119.

78. McDermott, "Prospect Theory," 293.

79. Meade, vol. 2, 134.

80. Robert Bauman and Lawrence Yates, "*My Clan Against the World*": US and Coalition Forces in Somalia,

1992–1994 (Fort Leavenworth: Combat Studies Institute Press, 2003), 101.

81. Mark Bowden, *Black Hawk Down: A Story of Modern War* (New York: Grove Press, 2010), 310.

82. Robert Oakley, PBS *FRONTLINE* interview, "Ambush in Mogadishu," http://www.pbs.org/wgbh/pages/frontline/shows/ambush/interviews/oakley.html (accessed February 11, 2012).

83. Baumann, 169.

84. Sam Kiley, "Will Aidid Play Ball?" *U. S. News & World Report*, October 18, 1993, 25.

85. Bowden, 311.

86. *United States Forces, Somalia After Action Report and Historical Overview: The United States Army in Somalia, 1992–1994* (Washington, DC: Center for Military History, 2003), 108.

87. Bauman, 169.

88. Bowden, 311.

89. Bauman, 168–169.

90. Anthony Zinni, PBS *FRONTLINE* interview, "Ambush in Mogadishu," *http://www.pbs.org/wgbh/pages/frontline/shows/ambush/interviews/zinni.html* (accessed February 11, 2012.

91. McDermott, "Prospect Theory," 300. See also Robert Jervis, "Political Implications of Loss Aversion," *Political Psychology* 13, no. 2 (June 1992), 190–191.

92. McDermott, "Prospect Theory," 300.

93. Michael Elliott, "The Making of a Fiasco," *Newsweek*, October 18, 1993, 34.

94. William Clinton, "Speech to the Nation: Somalia," October 7, 1993.

95. Ibid.

96. Alex Mintz and Karl DeRouen, *Understanding Foreign Policy Decision Making* (Cambridge: Cambridge University Press, 2010), 150.

97. Clinton, speech.

98. McDermott, "Prospect Theory," 300.

99. Kenneth Walsh, "The Unmaking of Foreign Policy," *U. S. News & World Report*, October 18, 1993, 32.

100. Jeffrey Berejikian, "Model Building with Prospect Theory: A Cognitive Approach to International Relations," *Political Psychology* 23, no. 4 (December 2002), 765.

101. Jervis, 191–192.

102. David Rieff, "A New Age of Imperialism?" *World Policy Journal* 16, no. 2 (Summer 1999): 5.

103. George Church, "Anatomy of a Disaster," *Time*, October 18, 1993, 49.

104. Jervis, 191.

105. Bowden, 311.

106. McDermott, "Prospect Theory," 296–297 notes this advantage of prospect theory. See Jervis, 191–192 for a similarly nuanced analysis of the Ford Administration's 1976 response to the swine flu epidemic scare.

107. James Goldgeier and Philip Tetlock, "Psychological Approaches," in *The Oxford Handbook of International Relations*, ed. Christian Reus-Smit and Duncan Snidal (Oxford: Oxford University Press, 2010), 464–466.

108. Jonathan Mercer, "Prospect Theory and Political Science," *Annual Review of Political Science* 8, 3.

109. Goldgeier and Tetlock, 464.

110. Moshe Levy and Haim Levy, "Prospect Theory: Much Ado about Nothing?" *Management Science* 48, no. 10 (October 2002), 1347.

111. Goldgeier and Tetlock, 466.

112. Mercer, 17.

113. Peter Wakker, "The Data of Levy and Levy (2002) 'Prospect Theory: Much Ado about Nothing?' Actually Support Prospect," *Management Science* 49, no. 7 (July 2003), 981.

114. Jervis, 188.

115. McDermott, "Prospect Theory," 304.

116. Ibid., 310.

CHAPTER 3

1. Alex Mintz and Karl DeRouen, *Understanding Foreign Policy Decision Making* (Cambridge, UK: Cambridge University Press, 2010), 80.

2. Ibid., 79.

3. Alex Mintz, "Applied Decision Analysis: Utilizing Poliheuristic Theory to Explain and Predict Foreign Policy and National Security Decisions," *International Studies Perspectives* 6 (2005), 94; Russell Bova, *How the World Works: A Brief Survey of International Relations* (New York: Longman, 2010), 91; Karl DeRouen and Christopher Sprecher, "Initial Crisis Reaction and Poliheuristic Theory," *The Journal of Conflict Resolution* 48, no. 1, (February 2004), 66.

4. Mintz, "Applied Decision Analysis," 94.

5. Steven Reed, "The Influence of Advisors and decision Strategies on Foreign Policy Choices: President Clinton's Decision to Use Force in Kosovo," *International Studies Perspectives* 6 (2005), 132. See also, Mintz and DeRouen, 34.

6. Alex Mintz, "The Decision to Attack Iraq: A Noncompensatory Theory of Decision Making," *Journal of Conflict Resolution* 37, 598.

7. Reed, 145.

8. Mintz and DeRouen, 34–35.

9. Alex Mintz. "How Do Leaders Make Decisions?: A Poliheuristic Perspective." *The Journal of Conflict Resolution* 48, no. 1, (February 2004), 4.

10. Ibid., 6 and Mintz and DeRouen, 79. Mintz also notes "poli" also refers to the notion that political leaders measure gains and losses in political terms.

11. Mintz and DeRouen, 34 and 36.

12. Brian McCauley, "Hungary and Suez, 1956: The Limits of Soviet and American Power," *Journal of Contemporary History* 16, no. 4 (October 1981), 778, 782, and 796.

13. Quoted in Robert Divine, *Since 1945: Politics and Diplomacy in Recent American History* (New York: McGraw-Hill, Inc, 1985), 76.

14. James Marchio, "Risking General War in Pursuit of Limited Objectives: US Military Contingency Planning for Poland in the Wake of the 1956 Hungarian Uprising," *The Journal of Military History* 66, no. 3 (July 2002), 789.

15. Robert Murphy, *Diplomat Among Warriors: The unique world of a Foreign Service Expert* (Garden City, NY: Doubleday & Company, Inc, 1964), 429–430.

16. Quoted in McCauley, 779.

17. Robert Doughty, et al. *American Military History and the Evolution of Western Warfare* (Lexington, MA: D. C. Health and Company, 1996), 584.

18. Robert Lieber, *No Common Power: Understanding International Relations* (New York: Harper Collins, 1991), 63–64.

19. Marchio, 791.

20. Murphy, 429–430.

21. Marchio, 791 and Murphy. 430.

22. Lieber, 64.

23. Murphy, 430.

24. Marchio, 792.

25. McCauley, 780 and Lieber, 61.

26. Murphy, 430.

27. Marchio, 790.

28. Murphy, 431.

29. "Jan Svoboda's Notes on the CPSU CC Presidium Meeting with Satellite Leaders, October 24, 1956," in *The*

1956 Hungarian Revolution: A History in Documents, ed. Csaba Békés (Budapest, HU: CEU Central European University Press, 2002), 225.

30. Quoted in McCauley, 783.

31. McCauley, 783.

32. Ibid., 789–790.

33. Resolution 1004 (ES-II), The Situation in Hungary, *http://www.un.org/arabic/documents/GADocs/A_3355english.pdf* (accessed December 4, 2012).

34. McCauley, 794.

35. Murphy, 431.

36. Quoted in Marchio, 783–784.

37. Quoted in Walter LaFeber, *America, Russia, and the Cold War* (New York: Alfred A. Knopf, 1985), 192.

38. Marchio, 810 and Doughty, 593.

39. Mintz and DeRouen, 36 and Ravi Dhar, "The Effect of Decision Strategy on Deciding to Defer Choice," *Journal of Behavioral Decision Making* 9 (1996): 268.

40. DeRouen and Sprecher, 58.

41. Karl DeRouen, "Foreign Policy Decision Making and the US Use of Force: Dien Bien Phu, 1954 and Grenada, 1983," presentation at the Annual Meeting of the Caribbean Studies Association, Kingston, Jamaica, May, 1993, 7 and 12. *http://ufdcimages.uflib.ufl.edu/CA/00/40/01/23/00001/PDF.pdf* (accessed November 5, 2012).

42. Manuel Falcon, "Bay of Pigs and Cuban Missile Crisis: Presidential Decision-making and its Effect on Military Employment during the Kennedy Administration" (Fort Leavenworth, KS: US Army Command and General Staff College, 1993), 24.

43. Lucien Vandenbroucke, "Anatomy of a Failure: The Decision to Land at the Bay of Pigs," *Political Science Quarterly* 99, no. 3 (Fall 1984), 473

44. Patrick James and Enhu Zhang, "Chinese Choices: A Poliheuristic Analysis of Foreign Policy Crises, 1950–1996," *Foreign Policy Analysis*, no. 1 (2005) 34.

45. Arthur Schlesinger, *A Thousand Days* (Boston, MA: Houghton Mifflin Co., 1965), 258.

46. DeRouen and Sprecher, 59.

47. Schlesinger, 249.

48. Thomas Paterson, et al, *American Foreign Policy: A History Since 1900* (Lexington, MA: D. C. Heath and Company, 1991), 540; Michael Grow, *US Presidents and Latin American Interventions: Pursuing Regime Change in the Cold War* (Lawrence: University Press of Kansas, 2008), 47–48.

49. Grayson Lynch, *Decision for Disaster: Betrayal at the Bay of Pigs* (Washington, DC: Brassey's, 1998), 14. See also Mintz and DeRouen, 107–108 for a discussion of Guatemala as a decision-making analog.

50. Dwight Eisenhower, *The White House Years: Waging Peace, 1956–1961* (Garden City, NY: Doubleday & Company, 1965), 401.

51. Juan Rodriguez, *The Bay of Pigs and the CIA* (New York: Ocean Press, 1999), 21; Peter Wyden, *Bay of Pigs: The Untold Story* (New York: Simon and Schuster, 1979), 24–25.

52. Grow, 49.

53. Ibid., 52.

54. Inaugural Address, John F. Kennedy, January 20, 1961, *http://www.jfklibrary.org/Asset-Viewer/BqXIEM9F4024ntFl7SVAjA.aspx* (accessed November 10, 2012).

55. Vandenbroucke, 484.

56. Schlesinger, 256.

57. Quoted in Divine, 106.

58. Grow, 53.

59. Divine, 106.

60. Quoted in Grow, 53.

61. Vandenbroucke, 484 and 488.

62. Mintz and DeRouen, 33.

63. Vandenbroucke, 484.

64. Ibid., 478.

65. Ibid., 477 and 478.

66. Ibid., 473.

67. Schlesinger, 237–238, 242, and 243 and Lynch, 29–30..

68. Luis Aguilar, *Operation Zapata* (Frederick, MD: University Publications of America, 1981), 13.

69. Aguilar, 13–14.

70. Wyden, 170.

71. Ibid., 199.

72. Peter Kornbluth, *Bay of Pigs Declassified* (New York: The New Press, 1998), 305–306.

73. Wyden, 199; Richard Bissell, *Reflections of a Cold Warrior* (Princeton, NJ: Yale University Press, 1996), 183.

74. Sorenson, 301.

75. Ibid., 302.

76. Schlesinger, 243.

77. Ibid., 243.

78. Ibid., 295.

79. Ibid., 256.

80. Ibid., 257.

81. Vandenbroucke, 471.

82. Divine, 106.

83. Theodore Sorensen, *Kennedy* (New York: Harper and Row, 1965), 309.

84. Doughty, 657.

85. Lyndon Johnson *The Vantage Point: Perspectives of the Presidency, 1963–1969* (New York: Holt, Rinehart, and Winston, 1971), 417.

86. Johnson, 386.

87. William Westmoreland, *A Soldier Reports* (Garden City, NY: Doubleday & Company, Inc, 1976), 354–356.

88. Mintz and DeRouen, 87.

89. Mintz, "Applied Decision Analysis," 95–97.

90. Johnson, 389–390.

91. Mintz, "Applied Decision Analysis," 96.

92. Johnson, 392 and Herring, 157.

93. Guenther Lewy, *America in Vietnam* (New York: Oxford University Press, 1978), 132

94. Lewy, 425.

95. Dave Palmer, *Summons of the Trumpet: US-Vietnam in Perspective* (San Rafael, CA: Presidio Press, 1978), 181 and Herring, 147.

96. Quoted in Don Oberdorfer, *Tet! The Turning Point in the Vietnam War* (Baltimore, MD: The Johns Hopkins University Press, 2001), 157.

97. Robert Dalleck, *Flawed Giant: Lyndon Johnson and His Times; 1961–1973* (New York: Oxford University Press, 1998), 506.

98. Johnson, 406–407. See also Johnson, 385.

99. Lewy, 432.

100. Dalleck, 515.

101. Johnson, 423.

102. Lewy, 129–130.

103. Johnson, 385, 391–392.

104. Quoted in Robert Strong, *Decisions and Dilemmas: Case Studies in Presidential Foreign Policy Making Since 1945* (Armonk, NY: M. E. Sharpe, 2005), 68.

105. George Ball, *The Discipline of Power: Essentials of a Modern World Structure* (London, UK: Bodley Head, Ltd, 1968), 334.

106. Quoted in Strong, 84.

107. Johnson, 388–389. Dalleck, 514 notes, "The North Korean action was subsequently seen as tied to Hanoi's Tet offensive."

108. Johnson, 385.

109. Ibid., 389.

110. Quoted in Dalleck, 507 and Herring, 156.

111. Oberdorfer, 159; Dalleck, 507.
112. Lewy, 434.
113. Dalleck, 506–507.
114. Mintz, 94.
115. Palmer, 206.
116. Reed, 132.
117. Palmer, 207.
118. Westmoreland, 353.
119. Quoted in Dalleck, 512.
120. Dalleck, 509.
121. Palmer, 205.
122. Mintz, "Applied Decision Analysis," 97.
123. Oberdorfer, 287–288.
124. Lewy, 165.
125. Herring, 151.
126. Quoted in Herring, 158.
127. Dalleck, 460. See also Herring, 152.
128. Westmoreland, 410.
129. Ibid., 359.
130. Dalleck, 512.
131. Kevin Dougherty and Jason Stewart, *The Timeline of the Vietnam War* (San Diego, CA: Thunder Bay Press, 2008), 171.
132. Herring, 152.
133. Ibid., 163–164.
134. Ibid., 152.
135. Mintz. "How Do Leaders Make Decisions?," 7–8.
136. Mintz and DeRouen, 90.
137. Ibid., 18.
138. Mintz. "How Do Leaders Make Decisions?," 10.
139. Mintz and DeRouen, 18.
140. David Brule, "The Poliheuristic Research Program: An Assessment and Suggestions for Further Progress," *International Studies Review*, no. 10, (2008), 283.

CHAPTER 4

1. Joshua Goldstein and Jon Pevehouse, *International Relations* (New York: Longman, 2010), 106.
2. Graham Allison, *Essence of Decision: Explaining the Cuban Missile Crisis* (Boston, MA: Little, Brown, 1971), 176.
3. Russell Bova, *How the World Works: A Brief Survey of International Relations* (New York: Longman, 2010), 94.
4. Alex Mintz and Karl DeRouen, *Understanding Foreign Policy Decision Making* (New York: Cambridge University Press, 2010), 71.
5. Archer Jones, *Confederate Strategy from Shiloh to Vicksburg* (Baton Rouge: Louisiana State University Press, 1961), 25–26. Hereafter Jones, *Confederate Strategy.*
6. Jones, *Confederate Strategy,* 25–26; Thomas Connelly and Archer Jones, *The Politics of Command: Factions and Ideas in Confederate Strategy* (Baton Rouge: Louisiana State University Press, 1973), 89.
7. Connelly and Jones, 189.
8. Robert Garlick Hill Kean, *Inside the Confederate Government: The Diary of Robert Garlick Hill Kean,* ed. Edward Younger (New York: Oxford University Press, 1957), 167 and 80.
9. Connelly and Jones, 171.
10. Mintz and DeRouen, 73.
11. Connelly and Jones, xv. See also Ibid., 88.
12. Jones, *Confederate Strategy,* 26.
13. Eugene Wittkopf, et al. *American Foreign Policy: Pattern and Process* (Belmont, CA: Wadsworth Publishing, 2007), 484.
14. Kean, 72. See also Ibid., 101.
15. Connelly and Jones, 197 and 200.
16. Mintz and DeRouen, 71.
17. Connelly and Jones, 121.
18. Jones, *Confederate Strategy,* 211.
19. *Official Records,* XXV, Pt. 2, 790.
20. Connelly and Jones, 42.
21. Mintz and DeRouen, 71.
22. Graham Allison, "Conceptual Models and the Cuban Missile Crisis," *The American Political Science Review* 63, No. 3 (September 1969), 710.
23. Ibid., 707.
24. Jones, *Confederate Strategy,* 233.
25. *Official Records,* XXV, Pt. 2, 713.
26. Connelly and Jones, 46.
27. Doughty, Robert, et al. *American Military History and the Evolution of Western Warfare* (Lexington, MA: D. C. Heath and Company, 1996), 106–107.
28. Connelly and Jones, 119.
29. Archer Jones, *Civil War Command & Strategy* (New York: The Free Press, 1992), 123. Hereafter, Jones, *Command & Strategy.*
30. Connelly and Jones, 184.
31. Doughty, 165.
32. Jones, *Command & Strategy,* 124.
33. Connelly and Jones, 173.
34. Ibid., 189.
35. Wittkopf, 484.
36. Kean, 80.
37. Ibid., 72.
38. Kevin Dougherty, *The Campaigns for Vicksburg, 1862–1863: Leadership Lessons* (Philadelphia, PA: Casemate, 2011), 52.
39. Bova, 95.
40. Connelly, 47. See also Ibid., 36–48.
41. Bova, 94.
42. Bruce Palmer, Jr., *The 25 Year War: America's Military Role in Vietnam* (Lexington: The University Press of Kentucky, 1984), 27.
43. Billy Mossman, *United States Army in the Korean War: Ebb and Flow, November 1950-July 1951* (Washington, DC: Center of Military History, United States Army, 1990), 32.
44. Elizabeth Schafer, "Helicopters in the Korean War," *The Korean War: An Encyclopedia,* ed. Stanley Sandler (New York: Garland Publishing, Inc, 1995), 130.
45. James M. Gavin, "Cavalry, and I Don't Mean Horses!" *Armor* 67, no. 3 (May/June 1958), 18.
46. Ian Horwood, *Interservice Rivalry and Airpower in the Vietnam* War (Fort Leavenworth, KS: Combat Studies Institute Press, 2006), 15–16.
47. Quoted in Rory Stewart and Gerald Kraus, *Can Intervention Work?* (New York: W. W. Norton, 2011), xxi.
48. John Tolson, *Airmobility, 1961–1971, Vietnam Studies* (Washington: Department of the Army, 1973), 11.
49. Horwood, 17.
50. Tolson, 11.
51. Shelby Stanton, *Anatomy of a Division: 1st Cav in Vietnam* (Novato, CA.: Presidio, 1987), 15–17.
52. Robert McNamara, Memorandum for Mr. Stahr, April 19, 1962 quoted in J. A. Stockfish, *The 1962 Howze Board and Army Combat Developments* (Santa Monica, CA: Arroyo Center, RAND, 1994), 41.
53. Tolson, 20.
54. Stanton, 17.
55. Kevin Dougherty, "The Evolution of Air Assault," *Joint Forces Quarterly* (Summer 1999), 54
56. Horwood, 123–124.
57. Ibid., 124.
58. Mark Olinger, *Conceptual Underpinnings of the Air Assault Concept: The Hogaboom, Rogers and Howze Boards*

(Arlington, VA: AUSA Institute of Land Warfare, 2006), 6.

59. Andrew Krepinevich, Jr., *The Army and Vietnam* (Baltimore, MD: The Johns Hopkins University Press, 1986), 118–24.

60. Tolson, 73.

61. Krepinevich, 124.

62. John Schlight, *Help From Above: Air Force Close Air Support of the Army1946–1973* (Washington, DC: Air Force History and Museums Program, 2003), 299.

63. Krepinevich, 124.

64. Palmer, 27.

65. Krepinevich, 124.

66. Horwood, 21.

67. Ibid., 23.

68. Ibid., 31.

69. Stockfish, 21.

70. Horwood, 43.

71. Stockfish, 21.

72. Ibid., 25.

73. Ibid., 25.

74. Horwood, 44.

75. Tolson, 105.

76. Horwood, 44.

77. Quoted in Lewis Sorley, *Honorable Warrior: General Harold K. Johnson and the Ethics of Command* (Lawrence: University Press of Kansas, 1998), 279.

78. Horwood, 132.

79. Ibid., 133; Sorley, 338.

80. Horwood, 133–134.

81. Douglas Macgregor, *Breaking the Phalanx: A New Design for Landpower in the 21st Century* (Westport, CN: Praeger, 1997), 51.

82. Ibid., 49.

83. John Blom, *Unmanned Aerial Systems: A Historical Perspective* (Fort Leavenworth, KS: Combat Studies Institute Press, 2010), 118–119.

84. Stew Magnumson, "Inter-Service Rivalry Surrounds Joint Heavy Lift Aircraft Program," *National Defense*, March 2009. *http://www.nationaldefensemagazine. org/archive/2009/March/Pages/Skepticism,Inter- ServiceRivalrySurroundsJointHeavyLiftAircraftProgram. aspx* (accessed July 23, 2012).

85. Horwood, 189.

86. See John Boykin, *Cursed Is the Peacemaker: The American Diplomat versus the Israeli General, Beirut 1982* (Belmont, CA: Applegate Press, 2002) for a detailed and sympathetic account of Habib's efforts.

87. Daniel Bolger, *Savage Peace: Americans at War in the 1990s* (Novato, CA: Presidio Press, 1995), 171; Caspar Weinberger, *Fighting for Peace* (New York: Warner Books, 1990), 143–144. Hereafter Weinberger, *Fighting*.

88. Boykin, 267.

89. Bolger, 172; Boykin, 267–269; Weinberger, *Fighting*, 150–151.

90. Eric Hammel, *The Root: The Marines in Lebanon, August 1982-February 1984* (St. Paul, MN: Zenith Press, 1999), 14.

91. Bolger, 172; Boykin, 271.

92. For a detailed discussion, see Peter Rodman, *Presidential Command: Power, Leadership and the Making of Foreign Policy from Richard Nixon to George W. Bush* (New York: Alfred A. Knopf, 2009).

93. Andrew Knight, "Ronald Reagan's Watershed Year?," *Foreign Affairs* 61, no. 3 (1982), 511.

94. Ibid., 513.

95. Ibid., 515.

96. Ibid., 532.

97. John Garofano, *The Intervention Debate: Towards a Posture of Principled Judgment* (Carlisle, PA: Strategic Studies Institute, 2002), 19.

98. George Shultz, "Low-Intensity Warfare: The Challenge of Ambiguity," Department of State, *Current Policy*, no. 783, January 1986.

99. George Shultz, *Turmoil and Triumph: My Years as Secretary of State* (New York: Scribner's, 1993), 350. Hereafter Shultz, *Turmoil and Triumph*.

100. Weinberger, *Fighting*, 159.

101. Ibid., 159.

102. Richard Hooker, "Presidential Decisionmaking and Use of Force: Case Study Grenada," *Parameters* xxi, no. 2 (Summer 1991), 62.

103. Mintz and DeRouen, 71.

104. Weinberger, *Fighting*, 159.

105. Ibid., 150–152.

106. Ibid., 154; Hammel, xxiii.

107. Weinberger, *Fighting*, 154.

108. Shultz, *Turmoil and Triumph*, 650.

109. Weinberger, *Fighting*, 155–156.

110. Wittkopf, 484.

111. Robert Jordan, "They Came in Peace," *Marine Corps Gazette*, July 1984, 57.

112. Weinberger, *Fighting*, 156.

113. Chris Lawson, "Peacekeeping Turned Sour 10 Years Ago, Too," *Army Times*, October 25, 1993, 11.

114. Weinberger, *Fighting*, 151.

115. Bolger, 174.

116. Ibid., 174.

117. Ibid., 182.

118. Jordan 57.

119. Weinberger, *Fighting*, 152.

120. Ibid., 151–152.

121. Shultz, *Turmoil and Triumph*.

122. Bova, 95.

123. Shultz, *Turmoil and Triumph*.

124. Caspar Weinberger, "The Uses of Military Power," *Defense 85*, January 1985, 2.

125. Goldstein and Pevehouse, 107; Bova, 95.

126. Allison, 690.

127. Mintz and DeRouen, 73.

CHAPTER 5

1. Russell Bova, *How the World Works* (New York: Longman, 2010), 96; Alex Mintz and Karl DeRouen, *Understanding Foreign Policy Decision Making* (New York: Cambridge University Press, 2010), 73.

2. Graham Allison, "Conceptual Models and the Cuban Missile Crisis," *The American Political Science Review* 63, no. 3 (September 1969), 702.

3. Dave Palmer, *Summons of the Trumpet* (San Rafael, CA: Presidio, 1978), 12.

4. Stephen Hosmer and George Tanham, "Countering Covert Aggression" (Santa Monica: RAND, January 1986), 17.

5. Anthony James Joes, *The War for South Viet Nam, 1954–197* (New York: Praeger, 2001), 64.

6. Guenter Lewy, *America in Vietnam* (New York: Oxford University Press, 1978), 112; Robert Doughty, *American Military History and the Evolution of Western Warfare* (Lexington, MA: D. C. Heath and Company, 1996), 642–643; Stanley Karnow, *Vietnam: A History* (New York: Penguin, 1997), 272–273.

7. Lewy, 112.

8. Palmer, 221.

9. Andrew Krepinevich Jr., *The Army and Vietnam* (Baltimore, MD: Johns Hopkins University Press, 1986), 5.

10. Jeffrey Record, *The American Way of War: Cultural*

Barriers to Successful Counterinsurgency (CATO Institute Policy Analysis no. 577, 2006), 10. *http://www.cato.org/publications/policy-analysis/american-way-war-cultural-barriers-successful-counterinsurgency* (accessed Jan 12, 2013).

11. Russel Weigley, *The American Way of War: A History of United States Military Strategy and Policy* (New York: Macmillan Publishing Co., Inc, 1973), 464.

12. *Public Papers of the Presidents of the United States: John F. Kennedy ... 1962* (Washington, DC: Government Printing Office, 1963), 454.

13. Weigley, 457.

14. Ibid., 466.

15. Jack Levy and William Thompson, *Causes of War* (Hoboken, NJ: John Wiley & Sons, 2009), 166.

16. Weigley, 467.

17. Allison, 698.

18. Robert Komer, "Clear, Hold, and Rebuild," Army 20, no. 5 (May 1970): 19.

19. Robert Komer, "Pacification," Army 20, no. 5 (June 1970): 23.

20. Maxwell Taylor, *Swords and Plowshares* (New York: W. W. Norton, 1972), 340. For other commentary on the lack of security, see Herring, 159; Lewy, 89; Willoughby, 6; Thayer, 137; John Cleland, "Principle of the Objective and Vietnam," *Military Review* (July 1966): 86; Edwin Chamberlain, "Pacification," *Infantry* (November–December 1968): 32–39; Cable, 257.

21. Palmer, 164.

22. Lawrence Yates, "A Feather in their CAP? The Marines' Combat Action Program in Vietnam." in *US Marines and Irregular Warfare, 1898–2007: Anthology and Selected Bibliography,* edited by Stephen Evans, 147–157. Quantico, VA: Marine Corps University Press, 2008, 148.

23. Ibid., 148–149; Al Hemingway, *Our War Was Different: Marine Combined Action Platoons in Vietnam* (Annapolis, MD: Naval Institute Press, 1994), 178; Anthony James Joes, *Resisting Rebellion: the History and Politics of Counterinsurgency* (Lexington, KY: University Press of Kentucky, 2006), 115.

24. Yates, 149; Lewy, 116–117; Andrew Brittle, *US Army Counterinsurgency and Contingency Operations Doctrine, 1942–1976* (Washington, DC: Center of Military History, 2006), 399–400; Swenson, 28; R. E. Williamson, "A Briefing for Combined Action," *Marine Corps Gazette*, March 1968, 41–43.

25. William Westmoreland, *A Soldier Reports* (Garden City, NY: Doubleday and Company, 1976), 166.

26. Kevin Dougherty, *The United States Military in Limited War: Case Studies in Success and Failure, 1945–1999* (Jefferson, NC: McFarland & Company, 2012), 86.

27. Palmer, 76.

28. Mintz and DeRouen, 74.

29. Lewy, 419.

30. Palmer, 75–79; Lewy, 378–379.

31. Lewy, 374.

32. Quoted in Ibid., 393.

33. Doughty, 651–652 and 658; Kevin Dougherty and Jason Stewart, *The Timeline of the Vietnam War* (London, UK: Amber Books, 2008), 40–41. The size of the ground force also rose from 184,300 at the end of 1965 to 485,600 at the end of 1967.

34. Mintz and DeRouen, 74.

35. FM 6-22, *Army Leadership* (Washington, DC: Department of the Army, 2006), 12–10.

36. William Steele, "The Iranian Hostage Rescue Mission: A Case Study" (Fort Lesley McNair, Washington, DC: National War College, 1984), 12.

37. Rose McDermott, *Risk-Taking in International Politics* (Ann Arbor: The University of Michigan Press, 2001), 52

38. See Ibid and Rose McDermott, "Prospect Theory in International Relations: The Iranian Hostage Rescue Mission," *Political Psychology* 13, no 2 (June 1992): 237–263.

39. Gary Sick, *All Fall Down: America's Tragic Encounter With Iran* (New York: Random House, 1985), 300–301.

40. Steele, 11.

41. Charlie Beckwith, *Delta Force: The Army's Elite Counterterrorist Unit* (New York: HarperCollins, 2000), 257.

42. FM 3.04–113, *Utility and Cargo Helicopter Operations* (Washington, DC: Headquarters, Department of the Army, 2007), 3–91.

43. Steele, 11–12; Beckwith, 258; and [Iranian Hostage] Rescue Mission Report, August 1980, 27. *http://www.history.navy.mil/library/online/hollowayrpt.htm.* (accessed January 27, 2013).

44. Beckwith, 253–256.

45. David Martin and John Walcott, *Best Laid Plans: The Inside Story of America's War Against Terrorism* (New York: Harper & Row, 1988), 34.

46. John Valliere, "Disaster at Desert One: Catalyst for Change," *Parameters* (Autumn 1992), 79

47. Russell Bova, *How the World Works* (New York: Longman, 2010), 96.

48. Rescue Mission Report, 30; Cogan, 212.

49. Mark Bowden, *Guests of the Ayatollah* (New York: Atlantic Monthly Press, 2006), 450; Rescue Mission Report, 45.

50. Beckwith, 305.

51. Charles Cogan, "Desert One and Its Disorders," *The Journal of Military History* 67, no.1 (January 2003), 212–214.

52. "'I sat there and cried,' colonel recalls," Tallahassee *Democrat*, May 2, 1980, 1A.

53. Beckwith, 306.

54. James Kyle, *The Guts to Try: The Untold Story of the Iran Hostage Rescue Mission by the on-Scene Desert Commander* (New York: Ballantine Books, 2002), 326.

55. Zbigniew Brzezinski, *Power and Principle: Memoirs of the National Security Adviser, 1977–1981* (New York: Farrar, Strauss, Giroux, 1983), 498.

56. Beckwith, 306.

57. Steele, 41.

58. Rescue Mission Report, 43.

59. Ibid., 10; Cogan, 212–214.

60. Bova, 96.

61. Bowden, 455–456.

62. Bova, 96.

63. Quoted in Cogan 215–216

64. Matt Matthews, *The Posse Comitatus Act and the United States Army: A Historical Perspective* (Fort Leavenworth, KS: Combat Studies Institute, 2006), 48.

65. James Delk. "Military Assistance in Los Angeles," *Military Review*, (September 1992), 16–18 and Christopher Schnaubelt, "Lessons in Command and Control from the Los Angeles Riots," *Parameters*, (Summer 1997), 90–99.

66. FM 100–5, *Operations* (Washington, DC: Department of the Army, 1986), 175.

67. Ibid., 176.

68. James Delk, "MOUT: A Domestic Case Study—The 1992 Los Angeles Riots," in *The City's Many Faces: Proceedings of the RAND Arroyo-MCWL-J8 UWG Urban Operations Conference, April 13–14, 1999*, ed. Russell Glen (Santa Monica, CA: RAND, 2000), 110.

69. Matthews, 5.

70. Ibid., 30–31.

71. Delk, 17; Matthews, 55; Schnaubelt, 100–102.

72. Matthews, 52.

73. Ibid., 52.

74. Ibid., 54.

75. Schnaubelt, 101.

76. Matthews, 53.

77. Ibid., 55.

78. Ibid., 56–57.

79. Ibid., 57–58.

80. FM 100–19, *Domestic Support Operations* (Washington, DC: Department of the Army, 1993), 3–2.

81. Matthews, 58–59.

82. Ibid., 54–55.

83. Bova, 96.

84. Gordon Sullivan, "Hurricane Andrew: An After Action Report." *Army*, (January 1993), 16.

85. Ibid., 18; Joint Task Force Andrew (JTFA) After Action Report (Miami, FL: Federal Emergency Management Agency, 1992), JTF-A AAR Tab I 2. Hereafter JTF-A AAR.

86. JTF-A AAR, 10.

87. Ibid., Tab S 2 and Sullivan, 19.

88. JTF-A AAR, Tab S 2.

89. Sullivan, 19.

90. CALL Newsletter 93–6, "Disaster Assistance" (Fort Leavenworth, KS: Center for Army Lessons Learned, October 1993), VII-1; JTFA AAR Overall Executive Summary 12, Tab A 18, and Tab S 2.

91. JTF-A AAR, Tab A 19.

92. Sullivan, 21.

93. JTF-A AAR, Overall Executive Summary 12 and Tab S 2.

94. Allison, 698.

95. Mintz and DeRouen, 72.

96. Joshua Goldstein and Jon Pevehouse, *International Relations* (New York: Longman, 2010), 106.

97. Mintz and DeRouen, 73 and Bova, 96.

98. Allison, 702.

99. Ibid., 700.

CHAPTER 6

1. Robert Wendzel, *International Politics: Policymakers & Policymaking* (New York: John Wiley, 1981), 439.

2. Glenn Hastedt, *American Foreign Policy: Past, Present, Future* (Upper Saddle River, NJ: Pearson, 2009), 270.

3. Ibid., 271.

4. Barbara Kellerman, "Allison Redux: Three More Decision-making Models," *Polity* 15, no. 3 (Spring 1983), 353.

5. John Niven, *Gideon Welles: Lincoln's Secretary of the Navy* (New York: Oxford University Press, 1973), 358.

6. Herman Hattaway and Archer Jones, *How the North Won*, (Urbana: University of Illinois Press, 1983), 135 and Rowena Reed, *Combined Operations in the Civil War*, (Annapolis, MD: Naval Institute Press, 1978), 8.

7. Kevin Weddle, "The Blockade Board of 1861 and Union Naval Strategy," Civil War History 48, no. 2 (June 2002), 131–132. For a more specific study of Du Pont, see Kevin Weddle, *Lincoln's Tragic Admiral: The Life of Samuel Francis Du Pont*, (Charlottesville, VA: University of Virginia Press, 2005). For more on the influence of Du Pont's experience in Mexico see Kevin Dougherty, *Civil War Leadership and Mexican War Experience* (Jackson: The University Press of Mississippi, 2007).

8. Kellerman, 353.

9. Weddle, 132.

10. "Alexander Dallas Bache: Leader of American Science and Second Superintendent of the United States Coast Survey," *http://celebrating200years.noaa.gov/historymakers/bache/welcome.html#beginnings* (accessed January 31, 2013).

11. Ibid.; Weddle, 128.

12. Weddle, 127–130.

13. Kellerman, 353. See also Graham Allison, *Essence of Decision: Explaining the Cuban Missile Crisis* (Boston: Little, Brown, 1971), 176.

14. Hugh Slotten, *Patronage, Practice, and the Culture of American Science: Alexander Dallas Bache and the U. S. Coast Survey* (New York: Cambridge University Press, 1994), 110.

15. Weddle, 132.

16. Ibid., 133.

17. Ibid., 133.

18. Kellerman, 353.

19. Weddle, 135 and 140.

20. Welles to Du Pont, Bache, Davis and Barnard, June 25, 1861, *Confidential Letter Book of the Secretary of the Navy, Correspondence of the Secretary of the Navy*. Quoted in Weddle, 134.

21. Weddle, 134.

22. Ibid., 134–136.

23. *The War of the Rebellion: A Compilation of the Official Records of the Union and Confederate Armies*; Series 1, Volume 53: 64. Hereafter *OR*.

24. *OR*, Series 1, Vol 53: 72–73.

25. Weddle, 137 and Reed, 8–9.

26. *Official Records of the Union and Confederate Navies in the War of the Rebellion*. Series I, Volume 16: South Atlantic Blockading Squadron: 618–630; Weddle, 138; and Ivan Musicant, *Divided Waters: The Naval History of the Civil War*, (Edison, NJ: Castle Books, 1995), 63.

27. Weddle, 139.

28. Ibid., 125.

29. Ibid., 141–142.

30. Gustavus Fox, *Confidential Correspondence of Gustavus Vasa Fox*, ed. Robert Thompson and Richard Wainwright (New York: De Vinne Press, 1918), 156.

31. Samuel F. Du Pont to Sophie Du Pont, Jan. 25, 1863, Du Pont, Letters, 2: 379.

32. Weddle, 125.

33. Donald Stoker, *The Grand Design: Strategy and the US Civil War* (New York: Oxford University Press, 2010), 95.

34. Weddle, 142.

35. National Security Action Memorandum 196. *http://microsites.jfklibrary.org/cmc/oct22/doc2.html*. (accessed February 3, 2013).

36. See Chapter 3 for a discussion of the Bay of Pigs.

37. Irving Janis, *Victims of Groupthink* (Boston, MA: Houghton Mifflin Company, 1972), 142.

38. Irving Janis, *Groupthink: Psychological Studies of Policy Decisions and Fiascoes* (Boston, MA: Houghton Mifflin Company, 1982), 9. Hereafter Janis, *Groupthink*

39. Joshua Goldstein and Jon Pevehouse, *International Relations* (New York: Longman, 2010), 111.

40. Steven Hook, *U.S. Foreign Policy: The Paradox of World Power*, 2nd ed. (Washington, D.C.: CQ Press, 2007), 92.

41. Janis, *Groupthink*, 9–10.

42. Ibid.141–143.

43. Theodore Sorensen, *Kennedy* (New York: Harper and Row, 1965), 674.

44. Kellerman, 354–355.

45. Janis, *Groupthink,* 141.

46. Goldstein and Pevehouse, 112.

47. Kellerman, 355.

48. Janis, *Groupthink,* 141; David Gibson, *Talk at the Brink: Deliberation and Decision during the Cuban Missile Crisis* (Princeton, NJ: Princeton University Press, 2012), 7.

49. Deborah Strober and Gerald Strober, *Kennedy Presidency: An Oral History of the Era* (Washington, D.C: Brassey's, 2003), 382.

50. Strober and Strober, 379.

51. Kellerman, 355–356.

52. Robert F. Kennedy, *Thirteen Days A Memoir of the Cuban Missile Crisis* (Boston, MA: W. W. Norton & Company, 1999), 9.

53. Janis, *Groupthink,* 142.

54. Kennedy, 26–27.

55. Janis, *Groupthink,* 142.

56. Kellerman, 361.

57. Gibson, 164.

58. Robert Thompson, *The Missiles of October: The Declassified Story of John F. Kennedy and the Cuban Missile Crisis* (New York: Simon & Schuster, 1992), 186.

59. Hastedt, 286–287.

60. Janis, *Groupthink,* 156.

61. Ibid., 158.

62. Peter Schweizer, *Reagan's War: The Epic Story of His Forty-Year Struggle and Final Triumph over Communism* (New York: Doubleday, 2002), 155.

63. National Security Decision Directive 32, "US National Security Strategy," (Washington, DC: The White House May 20, 1982), 1.

64. James Scott, "Interbranch Rivalry and the Reagan Doctrine in Nicaragua," *Political Science Quarterly* 112, no. 2 (Summer 1997): 237.

65. Schweizer, 204.

66. Ibid., 204.

67. Ibid., 205; John Prados, *Safe for Democracy: The Secret Wars of the CIA* (New York: Ivan R. Dee, 2006), 542; and Ivan Molloy, *Rolling Back Revolution* (Sterling, VA: Pluto Press, 2001), 69.

68. Prados, 542.

69. National Security Decision Directive 124, "Central America: Promoting Democracy, Economic Improvement, and Peace" (Washington, DC: The White House, February 7, 1984), 2.

70. Ibid., 5.

71. Peter Kornbluth, *Nicaragua: The Price of Intervention* (Washington, DC: Institute of Policy Studies, 1987), 139.

72. Michael Grow, *US Presidents and Latin American Interventions: Pursuing Regime Change in the Cold War* (Lawrence: University Press of Kansas, 2008), 131–132.

73. Kornbluth, 139.

74. Ibid., 140.

75. Holly Sklar, *Washington's War on Nicaragua* (Boston, MA: South End Press, 1988), 146.

76. Kornbluth, 141.

77. Ibid., 144.

78. "Sandinista Offensive," Department of State Bulletin, May 1988, 75; Kreischer, 37; Peter Range, "The End Game in Nicaragua," *U. S. News and World Report,* March 28, 1988, 17.

79. Kornbluth, 152.

80. Ned Ennis, "Exercise Golden Pheasant," *Military Review* (March 1989), 26.

81. NSSD 124, 5.

82. Range, 16; Harry Anderson, "The Contras Under the Gun," *Newsweek,* March 28, 1988, 36; Zinti, 15.

83. Schweizer, 205.

84. Scott, 242–243.

85. National Security Decision Directive 17, "Cuba and Central America" (Washington, DC: The White House, January 4, 1982), 2.

86. Grow, 134.

87. Ibid., 134–135; Robert Kagan, *A Twilight Struggle: American Power and Nicaragua, 1977–1990* (New York: Free Press, 1996), 202.

88. Schweizer, 205.

89. "Currents: US vs Nicaragua," *US News & World Report,* 28 April 1986, 8.

90. Scott, 237.

91. Caspar Weinberger, *Fighting For Peace: Seven Critical Years in the Pentagon* (New York: Warner Books, 1990), 375.

92. "Report of the Congressional Committees Investigating the Iran-Contra Affair," 508.

93. Kellerman, 353.

94. Bob Woodward, *Veil: The Secret Wars of the CIA, 1981–1987* (New York: Simon and Schuster, 2005), 271.

95. Joel Brinkley, "An Ex-Ambassador Says US Ordered Aid for Contras," New York *Times,* May 3, 1987, *http://www.nytimes.com/1987/05/03/world/an-ex-ambassador-says-us-ordered-aid-for-contras.html.* (accessed January 30, 2013).

96. Ibid.

97. Ibid.

98. Goldstein and Pevehouse, 111.

99. *The Tower Commission Report: The Full Text of The President's Special Review Board* (New York: Bantam, 1987), 80.

100. Paul Kowert, *Groupthink Or Deadlock: When Do Leaders Learn from Their Advisors?* (Albany, New York: SUNY Press, 2002), 148.

101. Kellerman, 353.

102. *The Tower Commission Report,* 79–81.

103. Goldstein and Pevehouse, 110–111.

104. Ibid., 112.

Chapter 7

1. Thomas Dye, *The Irony of Democracy: An Uncommon Introduction to American Politics* (Belmont, CA: Wadsworth, 2011), 1.

2. Ibid., 4.

3. Glenn Hastedt, *American Foreign Policy: Past, Present, Future* (Upper Saddle River, NJ: Prentice Hall, 2000), 249.

4. Ibid., 19–20.

5. J. Christopher Kovats-Bernat, "Factional Terror, Paramilitarism and Civil War in Haiti: The View from Port-au-Prince, 1994–2004," *Anthropologica* 48, no. 1: 130.

6. Christopher Mitchell, "US Policy toward Haitian Boat People, 1972–93," *Annals of the American Academy of Political and Social Science* 534 (July 1994): 74; Richard Melanson, *American Foreign Policy since the Vietnam War: The Search for Consensus from Nixon to Clinton* (Armonk: NY: M. E. Sharpe, 1996), 262.

7. Mitchell, 75.

8. Ibid., 75.

9. Melanson, 262.

10. Robert Keohane and Joseph Nye, *Power and Independence,* 2nd ed. (Cambridge, MA: Center for International Affairs, Harvard University with Harper Collins Publishers, 1989), 6.

11. Hastedt, 266.

12. Ibid., 274.

13. Walter Kretchik, et al. *Invasion, Intervention, "Intervasion": A Concise History of the U. S. Army in Operation Uphold Democracy* (Fort Leavenworth, KS: US Army Command and General Staff College Press, 1998), 8.

14. Ralph Pezzullo, *Plunging Into Haiti: Clinton, Aristide, And the Defeat of Diplomacy* (Jackson: University Press of Mississippi, 2006), 242 and Steven Holmes, "With Persuasion and Muscle, Black Caucus Reshapes Haiti Policy," New York Times, July 14, 1994. *http://www.nytimes.com/1994/07/14/world/with-persuasion-and-muscle-black-caucus-reshapes-haiti-policy.html* (accessed December 21, 2012).

15. Brian Robertson, "Black Caucus Wields Considerable Clout, But to Whose Benefit?" *Insight* (Washington Times), September 26, 1994, 6–7.

16. Hastedt, 274.

17. Raymond Copson, *The Congressional Black Caucus and Foreign Policy* (Hauppauge, NY: Nova Science Pub, 2003), 38.

18. Melanson, 263.

19. Tom Masland, "Should We Invade Haiti?" *Newsweek*, July 18, 1994, 42.

20. Kretchik, 33; Melanson, 262–263.

21. Kretchik, 34.

22. Ibid., 34.

23. Kate Doyle, "Hollow Diplomacy in Haiti," *World Policy Journal* 11, no. 1 (Spring 1994): 54.

24. Kretchik, 34–35.

25. James Pulley, Stephen Epstein, and Robert Cronin, "JTF Haiti: A United Nations Foreign Internal Defense Mission," *Special Warfare*, July 1994, 3.

26. Ibid., 35.

27. Ibid., 35. See also Peter Riehm, "The USS *Harlan County* Affair," *Military Review* 77, no. 4 (July-August 1994): 31–32.

28. Kretchik, 36–37; Riehm, 32.

29. Riehm, 33; Kretchin, 38.

30. Riehm, 33; Kretchin, 38.

31. Riehm, 33; Kretchin, 39.

32. Riehm, 33; Kretchin, 38.

33. Riehm, 34–35; Kretchin, 40–41.

34. Riehm, 35; Kretchin, 41.

35. Richard Millet, "Panama and Haiti," in *US and Russian Policy Making with Respect to the Use of Force*, ed. Jeremey Azrael and Emil Payin (Santa Monica, CA: RAND Corporation, 1996), 154.

36. Melanson, 263.

37. Jeffrey Record, "A Note on Interests, Values, and the Use of Force," *Parameters* (Spring 2001), 16.

38. Michael Mandelbaum, "Foreign Policy as Social Work," *Foreign Affairs* 75, no. 1 (Jan.-Feb., 1996), 17.

39. Ibid., 22.

40. Ibid., 22.

41. Dye, 1–18.

42. Masland, 43.

43. Holmes.

44. Ibid.

45. Ibid.

46. Washington *Post*, March 24, 1994. Quoted in Raymond Compson, *The Congressional Black Caucus & Foreign Policy* (1971–2002), (Hauppauge, NY: Nova Publishers, 2003), 15.

47. Copson, 38.

48. "Randall Robinson on Hunger Strike Until Clinton Changes Policy Toward Haiti," *Jet* 85, no. 26 (May 2, 1994), 4–5.

49. Pezzulo, 255.

50. Ibid., 257.

51. Kretchik, 56.

52. Copson, 38.

53. Carol Horner, "Point Man On Haiti Bill Gray's Life Took An Unexpected Turn When President Clinton Asked Him To Be His Special Adviser On Haiti. He Couldn't Say No," Philadelphia *Inquirer*, June 2, 1994. *http://articles.philly.com/1994–06–02/living/25832794_1_-haiti-political-asylum-white-house.* (accessed December 21, 2012).

54. Holmes.

55. William Gray, "Press Briefing by William Gray, Special Advisor to the President on Haiti," July 5, 1994. *http://www.presidency.ucsb.edu/ws/index.php?pid=59916* (accessed December 27, 2012).

56. Bill Clinton, "The Possible Invasion of Haiti to Restore a Democratic Government," *Vital Speeches of the Day*, no. 24, October 1, 1994, 739–740.

57. Julie Johnette O'Neal, "US Intervention in Grenada, Panama, and Haiti: A Social Constructionist Perspective," (Monterey, CA: Naval Postgraduate School, 1995), 97.

58. Dye, 4.

59. Kretchik, 184.

60. Copson, 39.

61. James Dobbins, et al. *America's Role in Nation-Building: From Germany to Iraq*, (Santa Monica, CA: RAND Corporation, 2003), 80 and 83.

62. Dye, 15.

63. Quoted in Harry Summers, *New World Strategy: A Military Policy for America's Future* (New York: Simon & Schuster, 1995), 11.

64. "Unforgettable Pictures of the Year," *Time*, December 28, 1992.

65. Those who discount the influence of the media in Somalia include Jonathan Mermin who calls Somalia "the myth of a media driven policy." He concludes that "if television inspired American intervention in Somalia, it did so under the influence of governmental actors — a number of senators, a House committee, a presidential candidate and figures within the Bush administration" (Hulme 51–52). Steven Livingston believes "The great majority of Somalia coverage followed rather than preceded official action" (8). Simon Hulme assesses that in Somalia "there was little if any media influence on foreign policy" (90). Those who minimize the CNN effect's role in Somalia rely largely on quantitative data. Livingston tracked daily CNN coverage of Somalia from July to December 1992 in terms of stories and minutes and concludes that those periods of expanded "media attention *followed* official actions" (Livingston 8). Mermin's analysis focuses not so much on the amount of coverage but its timing, arguing that the three full network stories reported on January 5, February 7, and March 2 occurred five to seven months before President Bush's decision and therefore had little impact (391). Mermin likewise discounts the three later full stories that ran July 22, July 31, and August 13 by superimposing political events over the reporting timeline and arguing that it was events in Washington such as Senator Nancy Kassebaum's July 22 Congressional testimony in support of intervention, rather than media coverage, which influenced policy (391–394).

66. Helen Metz, ed., *Somalia: A Country Study* (Washington, DC: Headquarters, Dept. of the Army, 1993), xiv, xxx-xxxiv; Frederick Fleitz, *Peacekeeping Fiascoes of the 1990s: Causes, Solutions, and U. S. Interests* (Westport, CN: Praeger, 2002), 130–131; "UN-mandated Force Seeks to Halt Tragedy: Operation Restore Hope," *UN Chronicle*, March 1993, 14.

67. Caspar Weinberger, "The Uses of Military Power," *Defense 85*, January 1985, 2. Hereafter, Weinberger, "Power."

68. Ibid., 10.

69. Edwin Arnold, "The Use of Military Power in Pursuit of National Interests," in *Parameters* (Spring 94), 5–7. Several alternatives to the Weinberger Doctrine were put forward in the aftermath of the Cold War. President George Bush's Chairman of the Joint Chiefs of Staff, General Colin Powell, argued that force should be used only as a last resort, there should be a clear-cut military objective, that the military objective must be measurable, and that military force should only be used in an overwhelming fashion. While similar in most ways to the Weinberger Doctrine, Powell's criteria significantly omitted the requirement of vital interest. However, others such as President Bill Clinton's Secretary of Defense Les Aspin did not feel that Powell's criteria went far enough in recognizing the changed world environment. At particular issue was the idea of using the military only as a last resort. Aspin and others argued for a more activist role in what is known as the "limited objective school." The heart of this thinking is that the military can be applied in one place to compel an adversary to change his behavior elsewhere. President Bush was also one of those who viewed the military as one of the means available at any time to achieve national interests, not only the means of last resort. See Ibid., 8–9.

70. Ibid., 10–11.

71. Matthew Harmon, "The Media, Technology and United states Foreign Policy: A Re-examination of the 'CNN Effect'" *Swords & Ploughshares: A Journal of International Affairs* VIII, no. 2 (Spring 1999), 1.

72. Stephen Livingstone, "Clarifying the CNN Effect: An Examination of Media Effects According to Type of Military Intervention." The Joan Shorenstein Center for Press, Politics, and Public Policy, (June 1997), 2.

73. Dye, 73 and 111.

74. Bruce Jentleson, *American Foreign Policy: The Dynamics of Choice in the 21st Century* (New York: W. W. Norton & Co, 2003), 55.

75. Livingstone, 6.

76. Speech by Anthony Lake to the TransAfrica Forum, Washington, DC June 29, 1995, in *US Department of State Dispatch* 6, no. 27 (July 3, 1995): 539.

77. Livingstone, 6.

78. Michael Mandelbaum, "The Reluctance to Intervene," *Foreign Policy* 95 (Summer 1994), 10.

79. Nik Gowing. "Real Time Television Coverage of Armed Conflicts and Diplomatic Crises: Does It Pressure or Distort Foreign Policy Decisions?" (The Joan Shorenstein Center for Press, Politics, and Public Policy, 1994), 49.

80. Craig Hines, "Pity, Not U. S. Security, Motivated Use of GIs in Somalia, Bush Says," The Houston *Chronicle,* October 24, 1999.

81. David Pearce, *Wary Partners — Diplomats and the Media* (Washington, DC: Congressional Quarterly, 1995), 18.

82. Jeffrey Record, "A Note on Interests, Values, and the Use of Force." *Parameters,* (Spring 2001), 16.

83. Dye, 114.

84. Hastedt, 249.

85. Dye, 1 and 95.

86. Quoted in Kristen Lord, *The Perils And Promise of Global Transparency: Why the Information Revolution May Not Lead to Security, Democracy, Or Peace* (Albany: State University of New York Press, 2006), 73.

87. Lord, 73. See also Steven Livingstone, "Suffering in Silence: Media Coverage of War and Famine in the Sudan." in *From Massacres to Genocide: The Media, Public Policy, and Humanitarian Crises,* Robert Rotberg, ed. (Washington, DC: Brookings Institution Press, 1996), 68–89.

88. Hastedt, 274.

89. Russell Watson, "It's Our Fight Now," *Newsweek,* December 14, 1992, 31.

90. Donald Snow, *Peacekeeping, Peacemaking and Peace Enforcement: The US Role in the New International Order* (Carlisle, PA: US Army War College, 1993), 3.

91. S. L. Arnold, "Somalia: An Operation Other Than War," *Military Review* (December 1993), 35; *U. S. Army Forces, Somalia, 10th Mountain Division (LI), AAR Summary* (Fort Drum, NY: Headquarters, 10th Mountain Division, 1993), 25. Hereafter, *10th Mountain AAR.*

92. *10th Mountain AAR,* 1.

93. Kenneth Allard, *Somalia Operations: Lessons Learned* (Washington, DC: National Defense University Press, 1995), 28.

94. Oakley says, "My own personal estimate is that there must have been 1,500 to 2,000 Somalis killed and wounded that day." Oakley *FRONTLINE* interview.

95. Clinton speech.

96. Colin Powell, *My American Journey: An Autobiography* (New York: Random House, 1995), 588.

97. Robert Kaplan, *Balkan Ghosts: A Journey Through History* (New York: St. Martin's Press, 1993), 39–40.

98. *Ethnic Cleansing in Kosovo: An Accounting* (Washington, DC: US Department of State, 1999), 3.

99. "Kosovo/Operation Allied Force After-Action Report," Department of Defense, January 31, 2000, 6.

100. Steven Redd, "The Influence of Advisers and Decision Strategies on Foreign Policy Choices: President Clinton's Decision to Use Force in Kosovo," *International Studies Perspectives* 6 (2005): 133.

101. Ibid., 140.

102. Ibid., 141.

103. Ibid., 142.

104. William Cohen, *FRONTLINE* Interview. http://www.pbs.org/wgbh/pages/frontline/shows/kosovo/interviews/cohen.html. (accessed December 30, 2012).

105. David Bergen, "Keeping a sense of balance," *U. S. News & World Report,* May 3, 1999, 72.

106. Michael Hirsh, "Albright's Old World Ways." *Newsweek,* March 29, 1999: 32.

107. Madeline Albright, *"FRONTLINE* Interview," http://www.pbs.org/wgbh/pages/frontline/shows/kosovo/interviews/albright.html. (accessed December 30, 2012).

108. Cohen, *FRONTLINE.*

109. Redd, 142–143.

110. Tom Carver, "Madeleine Albright: Haunted by history," BBC News, April 9, 1999. http://news.bbc.co.uk/2/hi/special_report/1999/03/99/kosovo_strikes/315053.stm. (accessed December 30, 2012).

111. Redd, 143.

112. James Rubin, "Press Briefing on the Kosovo peace talks," February 20, 1999. http://directshield.com/index.php?q=uggc%3A%2F%2F1997–2001.fgngr.tbi%2Fjjj%2Fcbyvpl_erznexf%2F1999%2F990220_ehova.ugzy. (accessed December 30, 2012).

113. Redd, 143.

114. Ibid., 143.

115. Ibid., 142.

116. Walter Isaacson, "Madeline's War," *Time,* May 17, 1999, 26.

117. Isaacson is an example.

118. Albright, *FRONTLINE.*

119. Steven Erlanger, "Albright Warns Serbs on Kosovo Violence, New York *Times,* March 8, 1998. http://www.nytimes.com/1998/03/08/world/albright-warns-serbs-on-kosovo-violence.html. (accessed December 31, 2012).

120. Carver.

121. Thomas Lippman, "State Dept. Miscalculated on Kosovo," Washington *Post,* April 7, 1999, A1.

122. Ann Blackman, *Seasons of Her Life: A Biography of Madeline Korbel Albright* (New York: Simon and Schuster, 1999), 19–20.

123. Isaacson, 26.

124. Ibid., 29.

125. Ibid., 26.

126. "Kosovo/Operation Allied Force After-Action Report," xiii.

127. Ibid., 79–81.

128. Kevin Whitelaw, "The Balkan Crisis is Clinton's defining moment," *U. S. News & World Report*, April 12, 1999, 16.

129. Bergen, 72.

130. Kevin Dougherty, "Kosovo: Long-term Impacts of Short-term Policy," *FAO Journal*, http://www.faoa.org/kosovo31.html (accessed Oct 27, 2004).

131. Philip Smucker, "Why America May or May Not, Arm the Rebels," *U. S. News & World Report*, April 26, 1998, 28.

132. "Report: U. S. officials expect Kosovo independence." http://www.cnn.com/WORLD/europe/9909/24/kosovo.us, (accessed August 3, 2004).

133. Chris Hedges, "Kosovo's Next Masters?" *Foreign Affairs*, May/June 1999, 24.

134. "NATO Kosovo Force: Troop Numbers & Contributions." http://www.aco.nato.int/kfor/about-us/troop-numbers-contributions.aspx (accessed January 4, 2013).

135. Carl Hodge, "Woodrow Wilson in Our Time: NATO's Goals in Kosovo," *Parameters* (Spring 2001): 130.

136. Ibid., 128.

137. Ibid., 131.

138. Hirsh, 32.

139. Wesley Clark, *Waging Modern War* (New York: PublicAffairs, 2001), 253. With great melodrama and self-import, Clark adds Albright also said "Now they will turn on you." He refrained from asking her who "they" were, however, claiming "it was too painful."

140. Dye, 4.

141. See Davenport, David. "The New Diplomacy," *Policy Review*, (Dec 2002 and Jan 2003): 17–30 for an interesting application of this phenomenon concerning nongovernmental organizations.

142. See Dye, 14–18.

CHAPTER 8

1. Thomas Dye, *The Irony of Democracy: An Uncommon Introduction to American Politics* (Belmont, CA: Wadsworth, 2011), 10–12.

2. Ibid., 10–12; Glenn Hastedt, *American Foreign Policy: Past, Present, Future* (Upper Saddle River, NJ: Prentice Hall, 2000), 275.

3. Hastedt, 275.

4. Martin Goldstein, *America's Foreign Policy: Drift or Decision* (New York: Rowman & Littlefield Publishers, 1984), 158.

5. David Donald, "Died of Democracy," in *Why the North Won the Civil War*, ed by David Donald (New York: Collier Books 1962), 90.

6. Archer Jones, *Confederate Strategy* (Baton Rouge: Louisiana State University Press, 1961), 42–49 and Kevin Dougherty, *Encyclopedia of the Confederacy* (San Diego, CA: Thunder Bay Press, 2010), 73–74.

7. Richard Beringer, et al. *Why the South Lost the War* (Athens: University of Georgia Press, 1986), 244.

8. Bell Irvin Wiley, *The Life of Johnny Reb* (Garden City, NY: Doubleday & Company, 1971), 125; Archer Jones, *Civil War Command & Strategy* (New York: The Free Press, 1992), 78.

9. Emory Thomas, *The Confederate Nation: 1861–1865* (New York: Harper & Row, 1979), 152.

10. Beringer, 287.

11. Ibid., 286–287.

12. Ibid., 205–206.

13. Ibid., 224.

14. Rod Andrew, *Wade Hampton: Confederate Warrior to Southern Redeemer* (Chapel Hill: University of North Carolina Press, 2008), 84; Donald, 82–83.

15. Wiley, 125.

16. Ibid., 126–127 and Thomas, 154.

17. Thomas, 152–153.

18. *War of the Rebellion: Official Records of the Union and Confederate Armies.* 128 volumes, (Washington, DC: Government Printing Office, 1880–1901), series 1, vol III, XLII, 1144. Hereafter OR.

19. Wiley, 125; Beringer, 226.

20. Theodore Lowi, "The Public Philosophy: Interest-Group Liberalism," *The American Political Science Review* 61, no. 1 (March 1967), 24.

21. James Randall, *The Civil War and Reconstruction* (Boston, MA: D. C. Heath and Company, 1937), 355.

22. Thomas, 154.

23. Beringer, 225–226.

24. Randall, 673–674.

25. See Theodore Lowi, *The End of Liberalism: Ideology, Policy, and the Crisis of Public Authority* (New York: N. W. Norton, 1969 and Hastedt, 276.

26. OR, series 4, vol II, 694.

27. Dougherty, 278–279.

28. Beringer, 434.

29. See endnote 17.

30. Wiley, 125; Dougherty, 124; Thomas, 154.

31. Dougherty, 278–279.

32. Wiley, 125; Dougherty, 74..

33. OR, series 4, vol II, 695 and Randall, 354.

34. Randall, 354.

35. Ben Severance, *Portraits of Conflict: A Photographic History of Alabama in the Civil War* (Fayetteville: University of Arkansas Press, 2012), 238.

36. Wiley, 144.

37. Ibid., 135 and 210; Dougherty, 94–95.

38. Wiley, 142–143; Dougherty, 125–126.

39. Wiley, 145.

40. Thomas, 154–155; Dougherty, 74.

41. Beringer, 288.

42. Donald, 81.

43. Robert Doughty et al, *American Military History and the Evolution of Warfare in the Western World* (Lexington, MA: D. C. Heath and Company, 1996), 681–684.

44. Caspar Weinberger, *Fighting for Peace* (New York: Warner Books, Inc, 1990), 388–389.

45. Michael Gurley, "Operation Earnest Will," (Newport, RI: US Naval War College, 1995), 3.

46. Text of Iranian Letter to the United Nations, reprinted in the New York *Times*, July 19, 1988, A9.

47. Michael Selby, "Without Clear Objectives: Operation Ernest Will," (Newport, RI: Naval War College, 1997), 1.

48. Weinberger, 396–397.

49. Jim Webb, *A Time to Fight: Reclaiming a Fair and Just America* (New York: Broadway, 2009), 135–137 and Weinberger, 398 and 401.

50. Bernard Trainor, "U.S. Officers Troubled by Plan to Aid Gulf Ships," New York *Times*, June 29, 1987, A6.

51. US Congress, Senate, Committee on Armed Services. US Military Forces to Protect "Re-flagged" Kuwaiti Oil Tankers. Hearing, 100th Congress, 1st Session 1987, 93.

52. Margaret G. Wachenfeld, "Reflagging Kuwaiti

Tankers: A U. S. Response in the Persian Gulf," *Duke Law Journal* 1988, no. 1 (February 1988): 185–186.

53. Michael Rehg, "Application of Decision-Making Models to Foreign Policy: A Case Study of the Reflagging of Kuwaiti Oil Tankers" (Wright-Patterson Air Force Base, OH: Air University, 1990), 64.

54. US Congress, House, Committee on Merchant Marine and Fisheries. Kuwaiti Tankers. Hearing, 100th Congress, 1st Session, 1987, 109.

55. Ibid., 110.

56. Ibid., 113.

57. Ibid., 228.

58. Rehg, 74.

59. Rehg, 77.

60. US Congress, Senate, Committee on Armed Services. Persian Gulf. Report to the Majority Leader, United States Senate, 100th, Congress, 1st Session, 1987, 11.

61. Rehg, 76.

62. US Congress, House, Committee on Merchant Marine and Fisheries. Kuwaiti Tankers. Hearing, 100th Congress, 1st Session, 1987, 46.

63. Ibid., 55.

64. Rehg, 77.

65. Wachenfeld, 186 and Rehg, 63.

66. Rehg, 63.

67. Dorothy Collin, "It's Reagan Vs. Congress In Standoff," Chicago *Tribune*, August 2, 1987. *http://articles.chicagotribune.com/1987–08–02/news/8702260219_1_-democrats-and-president-reagan-republican-lame-duck-president-senate-republicans.* (accessed February 20, 2013).

68. Amos Jordan, et al., *American National Security Strategy* (Baltimore, MD: The Johns Hopkins University Press, 2009), 26.

69. Weinberger, 428.

70. Collin.

71. Joseph Dawson, *Commanders in Chief: Presidential Leadership in Modern Wars* (Lawrence: University Press of Kansas, 1993), 17 and 45–46.

72. Fausold and Shank, 207. See also Weinberger, 400–401.

73. Martin Fausold and Alan Shank, *The Constitution and the American Presidency* (Albany: SUNY Press, 1991), 206–207 and Lowry v. Reagan, 676 F. Supp. 333 (D.D.C. 1987). See also Ackerman, David. "War Powers Litigation Since the Enactment of the War Powers Resolution." CRS Report RL30352.

74. Fausold and Shank, 208.

75. Ibid., 208.

76. Ibid., 209.

77. Dobbins, 87–88. The term "Bosniac" refers to Bosnian Muslims.

78. Michael Mandelbaum, "Foreign Policy as Social Work," *Foreign Affairs* (January/February 1996), 23.

79. Ryan Hendrickson, "War Powers, Bosnia, and the 104th Congress," *Political Science Quarterly* 113, no. 2 (Summer, 1998), 242.

80. Edward Barnes, "Time to Keep The Promise," *Time*, October 30, 1995, 78.

81. Bill Clinton, "Why Bosnia Matters to America, Newsweek, November 13, 1995, 55 and Steven Roberts, "Will the smiles fade?" *US News & World Report*, December 11, 1995.

82. Hendrickson, 243.

83. Barry Schweid, "House digs in its heels on US role in Bosnia," Atlanta *Constitution*, October 31, 1995, A1.

84. Helen Dewar and Michael Dobbs, "House Votes to Bar Sending Troops to Bosnia Without Hill Approval," Washington *Post*, November 18, 1995, A22.

85. Hendrickson, 245–246.

86. Ibid., 248.

87. "Senators demand congressional approval for Bosnian peacekeepers," The Benning *Leader*, October 8, 1995, 3.

88. Hendrickson, 248.

89. Ibid., 242.

90. Paul Glastris, "Jumping the gun in Bosnia," *US News & World Report*, November 27, 1995, 53.

91. "Congress questions peace goal," Columbus *Ledger-Enquirer*, December 1, 1995; Roberts, 45.

92. Jordan, 121.

93. Barnes, 79.

94. Hendrickson, 253–254.

95. Roberts, 45.

96. William Banks and Jeffrey Straussman, " Defense Contingency Budgeting in the Post-Cold-War World," *Public Administration Review* 59, no. 2 (March-April 1999), 137.

97. Hendrickson, 252.

98. Jordan, 26.

99. Ibid, 121; Dawson 44–45.

100. Dayton Peace Accord, 1995, Annex IA.

101. Ibid.

102. John MacInnis, "Contrasts in Peacekeeping: The Experience of UNPROPOR and IFOR in the Former Yugoslavia, *Mediterranean Quarterly* 1.8 no. 2 (Spring 1997), 158

103. Robert Perito, *The American Experience with Police in Peace Operations* (Clementsport, Canada: Canadian Peacekeeping Press, 2002), 50–51.

104. Wesley Clark, *Waging Modern War* (New York: PublicAffairs, 2001), 92.

105. Donald Snow, *When America Fights: The Uses of US Military Force* (Washington, DC: CQ Press, 2000), 129–130.

106. "Statement of Secretary of Defense William J. Perry On the Deployment of U.S. Troops with the Bosnia Peace Implementation Force," House Committee on International Relations and House Committee on National Security , November 30, 1995.

107. Jordan, 119.

108. Mandelbaum, 25.

109. Roland Paris, *At War's End: Building Peace After Civil Conflict* (New York: Cambridge University Press, 2004), 100–101; Dobbins, 108.

110. William Johnsen. *US Participation in IFOR: A Marathon, Not a Sprint* (Carlisle, PA: Strategic Studies Institute, 1996), 4.

111. Paris, 100; Dobbins, 108.

112. "...And Staying a lot Longer in Bosnia," *U. S. News & World Report*, November 25, 1996, 18.

113. Johnsen, ix.

114. Richard Swain, *Army Command in Europe During the Time of Peace Operations: Tasks Confronting US-AREUR Commanders, 1994–2000* (Carlisle Barracks, PA: Strategic Studies Institute, 2003), 97.

115. "NATO Ends SFOR Mission," NATO Update, December 2, 2004. *http://www.nato.int/docu/update/2004/12-december/e1202a.htm* (accessed March 9, 2013).

116. Snow, 111–112.

117. Smith, 47.

118. Hastedt, 275.

119. David Davenport, "The New Diplomacy," *Policy Review* (December 2002 and January 2003), 18.

120. Smith, 47.

121. Dye, 11.

Bibliography

Aguilar, Luis. *Operation Zapata*. Frederick, MD: University Publications of America, 1981.

"Alexander Dallas Bache: Leader of American Science and Second Superintendent of the United States Coast Survey." http://celebrating200 years.noaa.gov/historymakers/bache/welcome. html#beginnings (accessed January 31, 2013).

Allison, Graham. "Conceptual Models and the Cuban Missile Crisis." *The American Political Science Review* 63, no. 3 (September 1969): 689–718.

_____. *Essence of Decision: Explaining the Cuban Missile Crisis*. Boston: Little, Brown, 1971.

"...And Staying a lot Longer in Bosnia." *U. S. News & World Report*, November 25, 1996, 18.

Anderson, Harry. "The Contras Under the Gun." *Newsweek*, March 28, 1988, 36.

Andrew, Rod. *Wade Hampton: Confederate Warrior to Southern Redeemer*. Chapel Hill: University of North Carolina Press, 2008.

Appleman, Roy, et al. *Okinawa: The Last Battle*. Washington, DC: Center of Military History, 1948.

Art, Robert. "The Four Functions of Force." In *International Politics: Enduring Concepts and Contemporary Issues*, edited by Robert Art and Robert Jervis, 131–138. New York: Pearson, 2009.

Ball, George. *The Discipline of Power: Essentials of a Modern World Structure*. London: Bodley Head, 1968.

Banks, William, and Jeffrey Straussman. "Defense Contingency Budgeting in the Post-Cold-War World." *Public Administration Review* 59, no. 2 (March-April 1999): 135–146.

Barnes, Edward. "Time to Keep the Promise." *Time*, October 30, 1995, 78.

Bauman, Robert, and Lawrence Yates. *"My Clan Against the World": US and Coalition Forces in Somalia, 1992–1994*. Fort Leavenworth: Combat Studies Institute Press, 2003.

Beckwith, Charlie. *Delta Force: The Army's Elite Counterterrorist Unit*. New York: HarperCollins, 2000.

Belknap, Margaret. "The CNN Effect: Strategic Enabler or Operational Risk?" *Parameters* (Autumn 2002): 100–114.

Benjamin, Charles. "Hooker's Appointment and Removal." In *Battles and Leaders of the Civil War*, vol. III, 239–243. New York: The Century Company, 1888.

Berejikian, Jeffrey. "Model Building with Prospect Theory: A Cognitive Approach to International Relations." *Political Psychology* 23, no. 4 (December 2002): 759–786.

Beringer, Richard, et al. *Why the South Lost the War*. Athens: University of Georgia Press, 1986.

Bernstein, Barton. "Secrets and Threats: Atomic Diplomacy and Soviet-American Antagonism." In *Major Problems in American Foreign Relations, Volume II: Since 1914*, edited by Thomas Paterson and Dennis Merrill, 265–284. Lexington, MA: D. C. Heath, 1995.

Birtle, Andrew. *U.S. Army Counterinsurgency and Counterinsurgency Operations Doctrine, 1942–1976*. Washington, DC: Center of Military History, 2006.

Bissell, Richard. *Reflections of a Cold Warrior*. New Haven: Yale University Press, 1996.

Blechman, Barry, and Stephen Kaplan. *Force Without War: US. Armed Forces as a Political Instrument*. Washington, DC: Brookings Institution, 1978.

Blom, John. *Unmanned Aerial Systems: A Historical Perspective*. Fort Leavenworth: Combat Studies Institute Press, 2010.

Bolger, Daniel. *Savage Peace: Americans at War in the 1990s*. Novato, CA: Presidio Press, 1995.

Bonk, David. *Trenton and Princeton 1776-77: Washington Crosses the Delaware*. Oxford: Osprey, 2009.

Bova, Russell. *How the World Works: A Brief Survey of International Relations*. New York: Longman, 2010.

Bowden, Mark. *Black Hawk Down: A Story of Modern War*. New York: Grove Press, 2010.

_____. *Guests of the Ayatollah*. New York: Atlantic Monthly Press, 2006.

Boykin, John. *Cursed Is the Peacemaker: The American Diplomat versus the Israeli General, Beirut 1982*. Belmont, CA: Applegate Press, 2002.

Braestrup, Peter. *Big Story: How the American Press*

and Television Reported and Interpreted the Crisis of Tet 1968 in Vietnam and Washington. Garden City, NY: Anchor, 1978.

Brinkley, Joel. "An Ex-Ambassador Says US Ordered Aid for Contras." New York *Times*, May 3, 1987, *http://www.nytimes.com/1987/05/03/world/an-ex-ambassador-says-us-ordered-aid-for-contras.html* (accessed January 30, 2013).

Browne, Blaine. "MANHATTAN Project." In *World War II in Europe: An Encyclopedia*, edited by David Zabecki, 112. New York: Garland, 1999.

Brule, David. "The Poliheuristic Research Program: An Assessment and Suggestions for Further Progress." *International Studies Review*, no. 10 (2008): 266–293.

Brzezinski, Zbigniew. *Power and Principle: Memoirs of the National Security Adviser, 1977–1981.* New York: Farrar, Strauss, Giroux, 1983.

Burlingame, Michael. *Abraham Lincoln: A Life*, vol. II. Baltimore: Johns Hopkins University Press, 2008.

_____, and John R. Turner Ettlinger, eds. *Inside Lincoln's White House: The Complete Civil War Diary of John Hay.* Carbondale: Southern Illinois University Press, 1999.

Bush, George H. W. *All the Best, George Bush: My Life in Letters and Other Writings.* New York: Scribner, 2013.

CALL Newsletter 93–6, "Disaster Assistance." Fort Leavenworth: Center for Army Lessons Learned, October 1993.

CALL Newsletter No. 93–7, "Civil Disturbance." Fort Leavenworth: Center for Army Lessons Learned, November 1993.

Cashman, Greg. *What Causes War? An Introduction to Theories of International Conflict.* New York: Lexington Books, 1993.

Catton, Bruce. *Glory Road.* New York: Doubleday, 1952.

Chase, Philander, ed. *The Papers of George Washington*, Revolutionary War Series 7. Charlottesville: University Press of Virginia, 1997.

Cheney, Dick. *In My Time: A Personal and Political Memoir.* New York: Threshold Editions, 2012.

Church, George. "Anatomy of a Disaster." *Time*, October 18, 1993, 40–50.

Clark, Wesley. *Waging Modern War.* New York: PublicAffairs, 2001.

Clausewitz, Carl von. *On War*, edited by Michael Howard and Peter Paret. Princeton: Princeton University Press, 1984.

Cleaves, Freeman. *Meade of Gettysburg.* Norman: University of Oklahoma Press, 1960.

Clinton, William. "Speech to the Nation: Somalia." October 7, 1993.

_____. "Why Bosnia Matters to America." *Newsweek*, November 13, 1995, 55.

"The CNN Effect: How 24-Hour News Coverage Affects Government Decisions and Public Opinion." The Brookings Institution. https://apps49.brookings.edu/comm/transcripts/20020123.htm (accessed February 11, 2013).

Cogan, Charles. "Desert One and Its Disorders." *The Journal of Military History* 67, no.1 (January 2003): 201–216.

Cohen, Eliot. "The Mystique of U. S. Air Power." *Foreign Affairs* 73, no. 1 (January/February 1994): 109–123.

Collected Works of Abraham Lincoln, edited by Roy Basler, 9 vols. New Brunswick, NJ: Rutgers University Press, 1955.

Collin, Dorothy. "It's Reagan Vs. Congress In Standoff." Chicago *Tribune*, August 2, 1987, http://articles.chicagotribune.com/1987–08–02/news/8702260219_1_democrats-and-president-reagan-republican-lame-duck-president-senate-republicans (accessed February 20, 2013).

Combs, Cynthia. *Terrorism in the Twenty-First Century.* Boston: Pearson, 2013.

Compson, Raymond. *The Congressional Black Caucus & Foreign Policy (1971–2002).* Hauppauge, NY: Nova, 2003.

"Congress questions peace goal." Columbus *Ledger-Enquirer*, December 1, 1995.

Connelly, Thomas, and Archer Jones. *The Politics of Command: Factions and Ideas in Confederate Strategy.* Baton Rouge: Louisiana State University Press, 1973.

"Currents: US vs Nicaragua." *US News & World Report*, April 28, 1986, 8.

Cyr, Arthur. "Atomic Bomb, Decision to Use against Japan." In *World War II in the Pacific: An Encyclopedia*, edited by Stanley Sandler, 92–95. New York: Garland, 2001.

Dalleck, Robert. *Flawed Giant: Lyndon Johnson and His Times; 1961–1973.* Oxford: Oxford University Press, 1998.

Davenport, David. "The New Diplomacy." *Policy Review* (December 2002 and January 2003): 17–30.

Dawson, Joseph. *Commanders in Chief: Presidential Leadership in Modern Wars.* Lawrence: University Press of Kansas, 1993.

Delk, James. "Military Assistance in Los Angeles." *Military Review* (September 1992): 13–19.

DeRouen, Karl. "Foreign Policy Decision Making and the US Use of Force: Dien Bien Phu, 1954 and Grenada, 1983." Presentation at the Annual Meeting of the Caribbean Studies Association, Kingston, Jamaica, May 1993, 7, 12, http://ufdcimages.uflib.ufl.edu/CA/00/40/01/23/00001/PDF.pdf.

_____, and Christopher Sprecher. "Initial Crisis Reaction and Poliheuristic Theory." *The Journal of Conflict Resolution* 48, no. 1 (February 2004): 56–68.

Dewar, Helen, and Michael Dobbs. "House Votes to Bar Sending Troops to Bosnia Without Hill Approval." Washington *Post*, November 18, 1995, A22.

Dhar, Ravi. "The Effect of Decision Strategy on Deciding to Defer Choice." *Journal of Behavioral Decision Making* 9 (1996): 265–281.

Divine, Robert. *Since 1945: Politics and Diplomacy in Recent American History.* New York: McGraw-Hill, 1985.

Donald, David. "Died of Democracy." In *Why the North Won the Civil War*, edited by David Donald, 79–90. New York: Collier Books, 1962.

Dotto, Peter. "Marines in Los Angeles." *Marine Corps Gazette* (October 1992), 54–58.

Dougherty, Kevin. *The Campaigns for Vicksburg, 1862–1863: Leadership Lessons*. Philadelphia: Casemate, 2011.

_____. *Civil War Leadership and Mexican War Experience*. Jackson: University Press of Mississippi, 2007.

_____. *Encyclopedia of the Confederacy*. San Diego: Thunder Bay Press, 2010.

_____. "The Evolution of Air Assault." *Joint Forces Quarterly* (Summer 1999): 51–58.

_____, and Jason Stewart. *The Timeline of the Vietnam War*. San Diego: Thunder Bay Press, 2008.

Doughty, Robert, et al. *American Military History and the Evolution of Western Warfare*. Lexington, MA: D. C. Heath, 1996.

Dye, Thomas. *The Irony of Democracy: An Uncommon Introduction to American Politics*. Belmont, CA: Wadsworth, 2011.

Eisenhower, Dwight. *Crusade in Europe*. Garden City, NY: Doubleday, 1948.

_____. *The White House Years: Waging Peace, 1956–1961*. Garden City, NY: Doubleday, 1965.

Elliott, Michael. "The Making of a Fiasco." *Newsweek*, October 18, 1993, 34–38.

Ennis, Ned. "Exercise Golden Pheasant." *Military Review* (March 1989): 20- 26.

Falcon, Manuel. "Bay of Pigs and Cuban Missile Crisis: Presidential Decision-making and its Effect on Military Employment during the Kennedy Administration." Fort Leavenworth: US Army Command and General Staff College, 1993.

Fausold, Martin, and Alan Shank. *The Constitution and the American Presidency*. Albany: SUNY Press, 1991.

Felicetti, Gary, and John Luce. "The *Posse Comitatus* Act: Liberation from the Lawyers." *Parameters* (Autumn 2004): 94–107.

Fisher, David. *Washington's Crossing*. New York: Oxford University Press, 2006.

Fitzpatrick, John, ed. *The Writings of George Washington*, 39 vols. Washington, DC: Government Printing Office, 1931–1944.

FM 3.04–113, *Utility and Cargo Helicopter Operations*. Washington, DC: Headquarters, Department of the Army, 2007.

FM 5-0, *Army Planning and Orders Production*. Washington, DC: Department of the Army, 2005.

FM 6-0, *Mission Command: Command and Control of Army Forces*. Washington, DC: Department of the Army, 2003.

FM 6-22, *Army Leadership*. Washington, DC: Department of the Army, 2006.

FM 100–19, *Domestic Support Operations*. Washington, DC: Department of the Army, 1993.

Fox, Gustavus. *Confidential Correspondence of Gustavus Vasa Fox*, edited by Robert Thompson and Richard Wainwright. New York: De Vinne Press, 1918.

Freeman, Douglas Southall. *George Washington: A Biography*, 5 vols. New York: Charles Scribner's Sons, 1948–1952.

Garofano, John. *The Intervention Debate: Towards a Posture of Principled Judgment*. Carlisle, PA: Strategic Studies Institute, 2002.

Gaskins, Susanne Teepe. "Groves, Leslie." In *World War II in Europe: An Encyclopedia*, edited by David Zabecki, 327–328. New York: Garland, 1999.

Gavin, James. "Cavalry, and I Don't Mean Horses." *Armor*, May–June 1954, 18–22.

George Washington Papers at the Library of Congress, 1741–1799. http://memory.loc.gov/cgi- bin/query/r?ammem/mgw:@field(DOCID+@lit(gw060307)).

Gibson, David. *Talk at the Brink: Deliberation and Decision during the Cuban Missile Crisis*. Princeton: Princeton University Press, 2012.

Glastris, Paul. "Jumping the gun in Bosnia." *US News & World Report*, November 27, 1995, 53.

Goldstein, Joshua, and Jon Pevehouse. *International Relations*. New York: Longman, 2010.

Goldstein, Martin. *America's Foreign Policy: Drift or Decision*. Wilmington, DE: Scholarly Resources, 1984.

Golgeier, James, and Philip Tetlock. "Psychological Approaches." In *The Oxford Handbook of International Relations*, edited by Christian Reus-Smit and Duncan Snidal, 462–480. NewYork: Oxford University Press, 2010.

Gowing, Nik. "Real Time Television Coverage of Armed Conflicts and Diplomatic Crises: Does it Pressure or Distort Foreign Policy Decisions?" The Joan Shorenstein Center for Press, Politics, and Public Policy, 1994.

Graybill, Lyn. "CNN Made Me Do (Not Do) It." *Sarai Reader* (2004): 170–183.

Grow, Michael. *US Presidents and Latin American Interventions: Pursuing Regime Change in the Cold War*. Lawrence: University Press of Kansas, 2008.

Gurley, Michael. "Operation Ernest Will." Newport, RI: US Naval War College, 1995.

Hamilton, Lee, and Daniel Inouye. "Report of the Congressional Committees Investigating the Iran-Contra Affair." Collingdale, PA: Diane, 1995.

Hammel, Eric. *The Root: The Marines in Lebanon, August 1982-February 1984*. St. Paul: Zenith Press, 1999.

Harmon, Matthew. "The Media, Technology and United States Foreign Policy: A Re-examination of the 'CNN Effect.'" *Swords & Ploughshares: A Journal of International Affairs* VIII, no. 2 (Spring 1999), http://www.american.edu/sis/students/sword/spring99/USFP.PDF.

Hasegawa, Tsuyoshi. *Racing the Enemy: Stalin, Truman, and the Surrender of Japan*. Cambridge: Belknap Press, 2005.

Hastedt, Glenn. *American Foreign Policy: Past, Present, Future*. Upper Saddle River, NJ: Prentice Hall, 2000.

Hattaway, Hermann, and Archer Jones. *How the North Won*. Urbana: University of Illinois Press, 1983.

Hedges, Chris. "Kosovo's Next Masters?" *Foreign Affairs* (May/June 1999): 24–42.

Hendrickson, Ryan. "War Powers, Bosnia, and the 104th Congress." *Political Science Quarterly* 113, no. 2 (Summer 1998): 214–258.

Herring, George. *LBJ and Vietnam: A Different Kind of War.* Austin: University of Texas Press, 1994.

Hines, Craig. "Pity, not U. S. Security, Motivated Use of GIs in Somalia, Bush Says." Houston *Chronicle,* October 24, 1999, A11.

Hirsh, Michael. "Albright's Old World Ways." *Newsweek,* March 29, 1999, 32–33.

Hodge, Carl. "Woodrow Wilson in Our Time: NATO's Goals in Kosovo." *Parameters* (Spring 2001): 125–135.

Hoke, Jacob. *The Great Invasion.* Dayton: W. J. Shuey, 1887. *Holy Bible,* NIV.

Hook, Stephen. *U.S. Foreign Policy: The Paradox of World Power,* 2d ed. Washington, DC: CQ Press, 2007.

Hooker, Richard. "Presidential Decisionmaking and Use of Force: Case Study Grenada." *Parameters* xxi, no. 2 (Summer 1991): 61–72.

Horwood, Ian. *Interservice Rivalry and Airpower in the Vietnam* War. Fort Leavenworth: Combat Studies Institute Press, 2006.

Hosmer, Stephen, and George Tanham. "Countering Covert Aggression." Santa Monica: RAND, January 1986.

Hulme, Simon. "The Modern Media: The Impact on Foreign Policy." Fort Leavenworth: Command and General Staff College, 2001.

[Iranian Hostage] Rescue Mission Report, August 1980, 27, *http://www.history.navy.mil/library/online/hollowayrpt.htm* (accessed January 27, 2013).

Isaacson, Walter. "'Madeleine's War." *Time,* May 17, 1999, 26–35.

James, D. Clayton. "Harry Truman: The Two War Chief." In *Commanders in Chief: Presidential Leaders in Modern Wars,* edited by Joseph Dawson, 107–126. Lawrence: University Press of Kansas, 1993.

James, Patrick, and Enhu Zhang. "Chinese Choices: A Poliheuristic Analysis of Foreign Policy Crises, 1950–1996." *Foreign Policy Analysis,* no. 1 (2005): 31–54.

"Jan Svoboda's Notes on the CPSU CC Presidium Meeting with Satellite Leaders, October 24, 1956." In *The 1956 Hungarian Revolution: A History in Documents,* edited by Csaba Békés, 222–227. Budapest: CEU Central European University Press, 2002.

Janis, Irving. *Groupthink: Psychological Studies of Policy Decisions and Fiascoes.* Boston: Houghton Mifflin, 1982.

_____. *Victims of Groupthink.* Boston: Houghton Mifflin, 1972.

Jenkins, Brian. "Defense Against Terrorism." *Political Science Quarterly* 101, no. 5 (1986): 773–786.

Jentleson, Bruce. *American Foreign Policy: The Dynamics of Choice in the 21st Century.* New York: W. W. Norton, 2003.

_____. "The Reagan Administration and Coercive Diplomacy: Restraining More Than Remaking." *Political Science Quarterly* 106, no. 1 (Spring 1991): 57–82.

Jervis, Robert. "Political Implications of Loss Aversion." *Political Psychology* 13, no. 2 (June 1992): 187–204.

Johnsen, William. *US Participation in IFOR: A Marathon, Not a Sprint.* Carlisle, PA: Strategic Studies Institute, 1996.

Joint Task Force Andrew (JTFA) After Action Report. Miami: Federal Emergency Management Agency, 1992.

Jones, Archer. *Civil War Command & Strategy.* New York: Free Press, 1992.

_____. *Confederate Strategy.* Baton Rouge: Louisiana State University Press, 1961.

Jordan, Amos, et al. *American National Security.* Baltimore: Johns Hopkins University Press, 2009.

Jordan, Robert. "They Came in Peace." *Marine Corps Gazette,* July 1984, 56–63.

Journals of the Continental Congress, Volume 6. Washington, DC: U.S. Government Printing Office, 1906.

Kagan, Robert. *A Twilight Struggle: American Power and Nicaragua, 1977–1990.* New York: Free Press, 1996.

Kahin, George, and John Lewis. *The United States in Vietnam.* New York: The Dial Press, 1967.

Kahnemasn, Daniel, and Amos Tversky. "Prospect Theory: An Analysis of Decision Under Risk." *Econometrica* 47, no. 2 (March 1979): 263–292.

Kanter, Arnold. "Intervention Decisionmaking in the Bush Administration." *Conference Report: U.S. and Russian Policymaking with Respect to the Use of Force,* edited by Jeremy R. Azrael and Emil A. Payin. RAND, CF-129-CRES, Santa Monica: CA: RAND, 1996. http://www.rand.org/publications /CF/CF129/CF-129.chapter10.html.

Kean, Robert Garlick Hill. *Inside the Confederate Government: The Diary of Robert Garlick Hill Kean,* edited by Edward Younger. Oxford: Oxford University Press, 1957.

Kellerman, Barbara. "Allison Redux: Three More Decision-making Models." *Polity* 15, no. 3 (Spring 1983): 351–367.

Kennedy, Robert. *Thirteen Days: A Memoir of the Cuban Missile Crisis.* New York: W. W. Norton & Company, 1999.

Keohane, Robert, and Joseph Nye. *Power and Independence,* 2d ed. Cambridge: Center for International Affairs, Harvard University with Harper-Collins, 1989.

Ketchum, Richard. *The Winter Soldiers: The Battles for Trenton and Princeton.* New York: Henry Holt, 1999.

Kiley, Sam. "Will Aidid play ball?" *U. S. News & World Report,* October 18, 1993, 34–35.

Knight, Andrew. "Ronald Reagan's Watershed Year?" *Foreign Affairs* 61, no. 3 (1982): 511–540.

Komer, Robert. "Clear, Hold, and Rebuild." *Army* 20, no. 5 (May 1970), 16–24.

_____. "Pacification." *Army* 20, no. 5 (June 1970), 20–29.

Kornbluth, Peter. *Bay of Pigs Declassified*. New York: The New Press, 1998.

_____. *Nicaragua: The Price of Intervention*. Washington, DC: Institute of Policy Studies, 1987.

Kowert, Paul. *Groupthink or Deadlock: When Do Leaders Learn from Their Advisors?* Albany: SUNY Press, 2002.

Krepinevich, Andrew, Jr. *The Army and Vietnam*. Baltimore: Johns Hopkins University Press, 1986.

Kydd, Andrew. "Methodological Individualism and Rational Choice." In *The Oxford Handbook of International Relations*, edited by Christian Reus-Smit and Duncan Snidal, 425–443. Oxford: Oxford University Press, 2010.

Kyle, James. *The Guts to Try: The Untold Story of the Iran Hostage Rescue Mission by the On-Scene Desert Commander*. New York: Ballantine, 2002.

LaFeber, Walter. *America, Russia, and the Cold War*. New York: Alfred A. Knopf, 1985.

Lawson, Chris. "Peacekeeping Turned Sour 10 Years Ago, Too." *Army Times*, October 25, 1993, 11.

Levy, Moshe, and Haim Levy. "Prospect Theory: Much Ado about Nothing?" *Management Science* 48, no. 10 (October 2002): 1334- 49.

Lewy, Guenther. *America in Vietnam*. Oxford: Oxford University Press, 1978.

Livingstone, Stephen. "Clarifying the CNN Effect: An Examination of Media Effects According to Type of Military Intervention." The Joan Shorenstein Center for Press, Politics, and Public Policy, June 1997.

Lowi, Theodore. *The End of Liberalism: Ideology, Policy, and the Crisis of Public Authority*. New York: N. W. Norton, 1969.

_____. "The Public Philosophy: Interest-Group Liberalism." *The American Political Science Review* 61, no. 1 (March 1967): 5–24.

Lynch, Grayson. *Decision for Disaster: Betrayal at the Bay of Pigs*. Washington, DC: Brassey's, 1998.

Macgregor, Douglas. *Breaking the Phalanx: A New Design for Landpower in the 21st Century*. Westport, CT: Praeger, 1997.

MacInnis, John. "Contrasts in Peacekeeping: The Experience of UNPROPOR and IFOR in the Former Yugoslavia." *Mediterranean Quarterly* 1.8, no. 2 (Spring 1997): 146–162.

Magnumson, Stew. "Inter-Service Rivalry Surrounds Joint Heavy Lift Aircraft Program." *National Defense*, March 2009.

Mandelbaum, Michael. "Foreign Policy as Social Work." *Foreign Affairs* (January/February 1996): 16–32.

_____. "The Reluctance to Intervene." *Foreign Policy* 95 (Summer 1994): 10.

Marchio, James. "Risking General War in Pursuit of Limited Objectives: US Military Contingency Planning for Poland in the Wake of the 1956 Hungarian Uprising." *The Journal of Military History* 66, no. 3 (July 2002):783–812.

Martin, David, and John Walcott. *Best Laid Plans: The Inside Story of America's War Against Terrorism*. New York: Harper & Row, 1988.

Matloff, Maurice. *American Military History*. Washington, DC: Office of the Chief of Military History, 1973.

Matthews, Matt. *The Posse Comitatus Act and the United States Army: A Historical Perspective*. Fort Leavenworth: Combat Studies Institute, 2006.

McCauley, Brian. "Hungary and Suez, 1956: The Limits of Soviet and American Power." *Journal of Contemporary History* 16, no. 4 (October 1981): 783–812.

McCullough, David. *Truman*. New York: Simon & Schuster, 1992.

McDermott, Rose. "Prospect Theory in Political Science: Gains and Losses from the First Decade." *Political Psychology* 25, no 2 (April 2004): 289–312.

_____. *Risk-Taking in International Politics: Prospect Theory in American Foreign Policy*. Ann Arbor: University of Michigan Press, 2001.

Meade, George. *The Life and Letters of George Gordon Meade Major-General United States Army*, vol. 2. New York: Charles Scribner's Sons, 1913.

Mercer, Jonathan. "Prospect Theory and Political Science." *Annual Review of Political Science* 8: 1–21.

Mermin, Jonathan. "Television News and American Intervention in Somalia: The Myth of a Media-Driven Foreign Policy." *Political Science Quarterly* 112: 385–403.

Millett, Allan, and Peter Maslowski. *For the Common Defense: A Military History of the United States of America*. New York: Free Press, 1984.

Mintz, Alex. "Applied Decision Analysis: Utilizing Poliheuristic Theory to Explain and Predict Foreign Policy and National Security Decisions." *International Studies Perspectives* 6 (2005): 94–98.

_____. "The Decision to Attack Iraq: A Noncompensatory Theory of Decision Making." *Journal of Conflict Resolution* 37 (1993): 595–618.

_____. "How Do Leaders Make Decisions? A Poliheuristic Perspective." *The Journal of Conflict Resolution* 48, no. 1 (February 2004): 3–13.

_____, and Karl DeRouen. *Understanding Foreign Policy Decision Making*. Cambridge: Cambridge University Press, 2010.

Molloy, Ivan. *Rolling Back Revolution*. Sterling, VA: Pluto Press, 2001.

Mossman, Billy. *United States Army in the Korean War: Ebb and Flow, November 1950-July 1951*. Washington, DC: Center of Military History, United States Army, 1990.

Murphy, Robert. *Diplomat Among Warriors: The Unique World of a Foreign Service Expert*. Garden City, NY: Doubleday, 1964.

National Security Action Memorandum 196. http://microsites.jfklibrary.org/cmc/oct22/doc2.html (accessed February 3, 2013).

National Security Decision Directive 17, "Cuba and Central America." Washington, DC: The White House, January 4, 1982.

National Security Decision Directive 32, "US National Security Strategy." Washington, DC: The White House, May 20, 1982.

National Security Decision Directive 124, "Central America: Promoting Democracy, Economic Improvement, and Peace." Washington, DC: The White House, February 7, 1984.

"NATO Ends SFOR Mission," NATO Update, December 2, 2004. http://www.nato.int/docu/update/2004/12-december/e1202a.htm (accessed March 9, 2013).

Neuman, Johanna. "The Media's Impact on International Affairs, Then and Now." *SAIS Review* no 16.1 (1996): 109–123.*http://muse.jhu.edu/journals/sais_review/v016/16.1neuman.html.*

Niven, John. *Gideon Welles: Lincoln's Secretary of the Navy.* Oxford: Oxford University Press, 1973.

Oakley, Robert. PBS FRONTLINE interview, "Ambush in Mogadishu." http://www.pbs.org/wgbh/pages/frontline/shows/ambush/interviews/oakley.html.

Oberdorfer, Don. *Tet! The Turning Point in the Vietnam War.* Baltimore: Johns Hopkins University Press, 2001.

Official Records of the Union and Confederate Navies in the War of the Rebellion. Washington, DC: Government Printing Office, 1894–1922.

Olinger, Mark. *Conceptual Underpinnings of the Air Assault Concept: The Hogaboom, Rogers and Howze Boards.* Arlington: AUSA Institute of Land Warfare, 2006.

Paine, Thomas. *The American Crisis.* London: James Watson, 1835.

Palmer, Bruce, Jr. *The 25 Year War: America's Military Role in Vietnam.* Lexington: University Press of Kentucky, 1984.

Palmer, Dave. *Summons of the Trumpet: US-Vietnam in Perspective.* San Rafael, CA: Presidio Press, 1978.

Papers of John Adams, vol. 5, Massachusetts Historical Society. http://www.masshist.org/publications/apde/portia.php?id=PJA05d034.

Paris, Roland. *At War's End: Building Peace After Civil Conflict.* Cambridge: Cambridge University Press, 2004.

Parks, W. Hays. "Crossing the Line." *Proceedings* (November 1986): 41–51.

Perito, Robert. *The American Experience with Police in Peace Operations.* Clementsport: Canadian Peacekeeping Press, 2002.

Persico, Joseph. *My Enemy, My Brother: Men and Days of Gettysburg.* New York: Da Capo Press, 1996.

Pezzullo, Ralph. *Plunging Into Haiti: Clinton, Aristide, and the Defeat of Diplomacy.* Jackson: University Press of Mississippi, 2006.

Powell, Colin. *My American Journey: An Autobiography.* New York: Random House, 1995.

Prados, John. *Safe for Democracy: The Secret Wars of the CIA.* Chicago: Ivan R. Dee, 2006.

Quillen, Chris. "*Posse Comitatus* and Nuclear Terrorism." *Parameters* (Spring 2002): 60–74.

Randall, James. *The Civil War and Reconstruction.* Boston: D. C. Heath, 1937.

"Randall Robinson on Hunger Strike Until Clinton Changes Policy Toward Haiti." *Jet* 85, no. 26 (May 2, 1994), 4–5.

Range, Peter. "The End Game in Nicaragua." *U. S. News and World Report,* March 28, 1988, 16–18.

Rapoport, David. "The Four Waves of Modern Terrorism." *http://www.international.ucla.edu/media/files/Rapoport-Four-Waves-of-Modern-Terrorism.pdf* (accessed March 11, 2013).

Record, Jeffrey. "A Note on Interests, Values, and the Use of Force." *Parameters* (Spring 2001): 15–21.

Redd, Steven. "The Influence of Advisors and decision Strategies on Foreign Policy Choices: President Clinton's Decision to Use Force in Kosovo." *International Studies Perspectives* 6 (2005): 129–150.

Reed, Rowena. *Combined Operations in the Civil War.* Annapolis: Naval Institute Press, 1978.

Rehg, Michael. "Application of Decision-Making Models to Foreign Policy: A Case Study of the Reflagging of Kuwaiti Oil Tankers." Wright-Patterson Air Force Base, OH: Air University, 1990.

Ricchiardi, Sherry. "Missed Signals." *American Journalism Review.* August/September 2004. http://www.ajr.org/Article.asp?id=3716 (accessed February 11, 2013).

Rieff, David. "A New Age of Imperialism?" *World Policy Journal* 16, no. 2 (Summer 1999): 1–10.

Roberts, Steven. "Will the smiles fade?" *US News & World Report,* December 11, 1995.

Robinson, Piers. "Theorizing the Influence of Media on World Politics: Models of Media Influence on Foreign Policy." *European Journal of Communication* 16, no. 4: 523–544.

Rodriguez, Juan. *The Bay of Pigs and the CIA.* New York: Ocean Press, 1999.

"Sandinista Offensive." Department of State Bulletin. May 1988, 73–74.

Sauers, Richard. *Meade: Victor of Gettysburg.* Washington, DC: Brassey's, 2003.

Scales, Robert. *Certain Victory: The US Army in the Gulf War.* Fort Leavenworth: Command and General Staff CollegePress, 1994.

Schafer, Elizabeth. "Helicopters in the Korean War." In *The Korean War: An Encyclopedia,* edited by Stanley Sandler, 129–131. New York: Garland, 1995.

Schlesinger, Arthur. *A Thousand Days.* Boston: Houghton Mifflin, 1965.

Schlight, John. *Help From Above: Air Force Close Air Support of the Army 1946–1973.* Washington, DC: Air Force History and Museums Program, 2003.

Schnaubelt, Christopher. "Lessons in Command and Control from the Los Angeles Riots." *Parameters* (Summer 1997): 88–109.

Schwarzkopf, Norman. *It Doesn't Take a Hero.* New York: Bantam Books, 1992.

Schweid, Barry. "House Digs in Its Heels on US Role in Bosnia." Atlanta *Constitution,* October 31, 1995, A1.

Schweizer, Peter. *Reagan's War: The Epic Story of His*

Forty-Year Struggle and Final Triumph over Communism. New York: Doubleday, 2002.

Scott, James. "Interbranch Rivalry and the Reagan Doctrine in Nicaragua." *Political Science Quarterly* 112, no. 2 (Summer 1997): 237–260.

Seib, Philip. "Politics of the Fourth Estate: The Interplay of Media and Politics in Foreign Policy." *Harvard International Review* 22, no. 3 (Fall 2000): 60.

"Senators Demand Congressional Approval for Bosnian Peacekeepers." The Benning *Leader*, October 8, 1995, 3.

"Serbia/Yugoslavia (4)." PollingReport.com. *http:// peoplepress.org/reports/print.php3?ReportID=64* (accessed February 11, 2013).

Severance, Ben. *Portraits of Conflict: A Photographic History of Alabama in the Civil War.* Fayetteville: University of Arkansas Press, 2012.

Shanahan, Jack. "The 'CNN Effect:' TV and Foreign Policy." http://www.cdi.org/adm/834/transcript. html (accessed February 11, 2013).

Sharkey, Jacqueline. "When Pictures Drive Foreign Policy." *American Journalism Review* 15, no. 10 (December 1993): 14–19.

Selby, Michael. "Without Clear Objectives: Operation Earnest Will." Newport, RI: Naval War College, 1997.

Sick, Gary. *All Fall Down: America's Tragic Encounter with Iran.* New York: Random House, 1985.

Skaggs, David. "The Generalship of George Washington." *Military Review* (July 1974): 3–10.

Sklar, Holly. *Washington's War on Nicaragua.* Boston: South End Press, 1988.

Slotten, Hugh. *Patronage, Practice, and the Culture of American Science: Alexander Dallas Bache and the U. S. Coast Survey.* Cambridge: Cambridge University Press, 1994.

Snow, Donald. *When America Fights: The Uses of US Military Force.* Washington, DC: CQ Press, 2000.

Sorensen, Theodore. *Kennedy.* New York: Harper & Row, 1965.

Sorley, Lewis. *Honorable Warrior: General Harold K. Johnson and the Ethics of Command.* Lawrence: University Press of Kansas, 1998.

Stackpole, Edward. *They Met at Gettysburg.* Mechanicsville, PA: Stackpole Books, 1982.

Stanik, Joseph. *El Dorado Canyon: Reagan's Undeclared War With Qaddafi.* Annapolis: Naval Institute Press, 2002.

Stanton, Shelby. *Anatomy of a Division: 1st Cav in Vietnam.* Novato, CA: Presidio Press, 1987.

"Statement of Secretary of Defense William J. Perry on the Deployment of U.S. Troops with the Bosnia Peace Implementation Force." House Committee on International Relations and House Committee on National Security, November 30, 1995. *http:// www.defense.gov/speeches/speech.aspx?speechid=102 6* (accessed March 12, 2013).

Steele, William. *The Iranian Hostage Rescue Mission: A Case Study.* Fort Lesley McNair, Washington, DC: National War College, 1984.

Strober, Deborah, and Gerald Strober. *Kennedy Pres-*

idency: An Oral History of the Era. Washington, D.C: Brassey's, 2003.

Stockfish, J. A. *The 1962 Howze Board and Army Combat Developments.* Santa Monica: Arroyo Center, RAND, 1994.

Stoker, Donald. *The Grand Design: Strategy and the US Civil War.* Oxford: Oxford University Press, 2010.

Strong, Robert. *Decisions and Dilemmas: Case Studies in Presidential Foreign Policy Making Since 1945.* Armonk, NY: M. E. Sharpe, 2005.

Sullivan, Gordon. "Hurricane Andrew: An After Action Report." *Army* (January 1993), 16–22.

Summers, Harry. *New World Strategy: A Military Policy for America's Future.* New York: Simon & Schuster, 1995.

_____. *Strategy: A Critical Analysis of the Vietnam War.* Novato, CA: Presidio Press, 1982.

Swain, Richard. *Army Command in Europe During the Time of Peace Operations: Tasks Confronting USAREUR Commanders, 1994–2000.* Carlisle Barracks, PA: Strategic Studies Institute, 2003.

Text of Iranian Letter to the United Nations, reprinted in the New York *Times*, July 19, 1988, A9.

Thomas, Emory. *The Confederate Nation: 1861–1865.* New York: Harper & Row, 1979.

Thompson, Robert. *The Missiles of October: The Declassified Story of John F. Kennedy and the Cuban Missile Crisis.* New York: Simon & Schuster, 1992.

Tolson, John. *Airmobility 1961–1971.* Washington, DC: Government Printing Office, 1989.

The Tower Commission Report: The Full Text of The President's Special Review Board. New York: Bantam, 1987.

Trainor, Bernard. "U.S. Officers Troubled by Plan to Aid Gulf Ships." New York *Times*, June 29, 1987, A6.

Trebilcock, Craig. "The Myth of *Posse Comitatus*." *Journal of Homeland Defense* (October 27, 2000): 1–5.

Truman, Harry. "Address Before a Joint Session of the Congress, April 16, 1945." *http://www.trumanlibrary.org/ww2/stofunio.htm* (accessed October 8, 2012).

_____. *Memoirs by Harry S. Truman*, vol 1. Garden City: Doubleday, 1955.

Tucker, David. "Facing the Facts: The Failure of Nation Assistance." *Parameters.* (Summer 1993): 34–40.

US Congress, House, Committee on Merchant Marine and Fisheries. Kuwaiti Tankers. Hearing, 100th Congress, 1st Session, 1987.

US Congress, Senate, Committee on Armed Services. U.S. Military Forces to Protect "Re-flagged" Kuwaiti Oil Tankers, Hearing, 100th Congress, 1st Session, 1987.

US Congress, Senate, Committee on Armed Services, Persian Gulf, Report to the Majority Leader, United States Senate, 100th Congress, 1st Session, 1987.

United States Forces, Somalia After Action Report and

Historical Overview: The United States Army in Somalia, 1992–1994. Washington, DC: Center for Military History, 2003.

Valliere, John. "Disaster at Desert One: Catalyst for Change." *Parameters* (Autumn 1992): 69–82.

Vandenbroucke, Lucien. "Anatomy of a Failure: The Decision to Land at the Bay of Pigs." *Political Science Quarterly* 99, no. 3 (Fall 1984): 471–491.

Wachenfeld, Margaret. "Reflagging Kuwaiti Tankers: A U. S. Response in the Persian Gulf." *Duke Law Journal* 1988, no. 1 (February 1988): 174–202.

Wakker, Peter. "The Data of Levy and Levy (2002) 'Prospect Theory: Much Ado about Nothing?' Actually Support Prospect." *Management Science* 49, no. 7 (July 2003): 979–981.

Walsh, Kenneth. "The Unmaking of Foreign Policy." *U. S. News & World Report*, October 18, 1993, 30–33.

Walt, Stephen. "International Relations: One World, Many Theories." *Foreign Policy* no. 110 (Spring 1998): 29–43.

War of the Rebellion: Official Records of the Union and Confederate Armies, 128 vols., Washington, DC: Government Printing Office, 1880–1901.

Webb, James. *A Time to Fight: Reclaiming a Fair and Just America*. New York: Broadway, 2009.

Weddle, Kevin. "The Blockade Board of 1861 and Union Naval Strategy." *Civil War History* 48, no. 2 (June 2002): 123–142.

_____. *Lincoln's Tragic Admiral: The Life of Samuel Francis Du Pont*. Charlottesville: University of Virginia Press, 2005.

Weigley, Russell. *History of the United States Army*. New York: Macmillan, 1967.

Weinberger, Caspar. *Fighting for Peace: Seven Critical Years in the Pentagon*. New York: Warner Books, 1990.

_____. "The Uses of Military Power." *Defense 85*, January 1985, 2–11.

Wendzel, Robert. *International Politics: Policymakers & Policymaking*. New York: John Wiley, 1981.

Westmoreland, William. *A Soldier Reports*. Garden City, NY: Doubleday, Inc, 1976.

Whitelaw, Kevin. "The Balkan Crisis Is Clinton's Defining Moment." *U. S. News & World Report*, April 12, 1999, 16–17.

Wiley, Bell Irvin. *The Life of Johnny Reb*. Garden City, NY: Doubleday, 1971.

Wittkopf, Eugene, et al. *American Foreign Policy: Pattern and Process*. Belmont, CA: Wadsworth, 2007.

Woodward, Bob. *Veil: The Secret Wars of the CIA, 1981–1987*. New York: Simon & Schuster, 2005.

Wombwell, James. *Army Support During Hurricane Katrina*. Fort Leavenworth: Combat Studies Institute Press, 2009.

Wyden, Peter. *Bay of Pigs: The Untold Story*. New York: Simon & Schuster, 1979.

Yarger, H. Richard, and George Barber. "The US Army War College Methodology for Determining Interests and Levels of Intensity." Carlisle Barracks, PA: US Army War College, 1997.

Yates, Lawrence. "A Feather in their CAP? The Marines' Combat Action Program in Vietnam." In *US Marines and Irregular Warfare, 1898–2007: Anthology and Selected Bibliography*, edited by Stephen Evans, 147–157. Quantico: Marine Corps University Press, 2008.

Zinni, Anthony. PBS FRONTLINE interview, "Ambush in Mogadishu." *http://www.pbs.org/wgbh/pages/frontline/shows/ambush/interviews/zinni.html* (accessed January 23, 2012).

Zinti, Robert. "The Contra Tangle." *Time*, March 28, 1988, 15.

Index